Stem Cells:
The Complete Blueprint

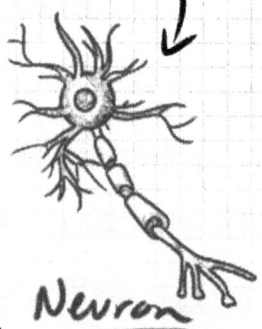

Blastocyst

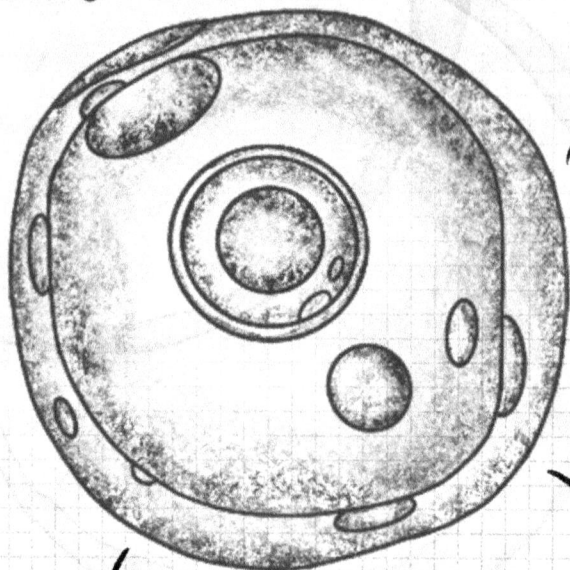

Stem Cell

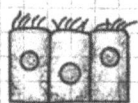

Fat cells

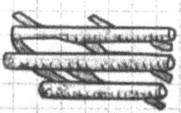

Cardiac cells

Enterocytes

Neuron

ii

Stem Cells: The Complete Blueprint

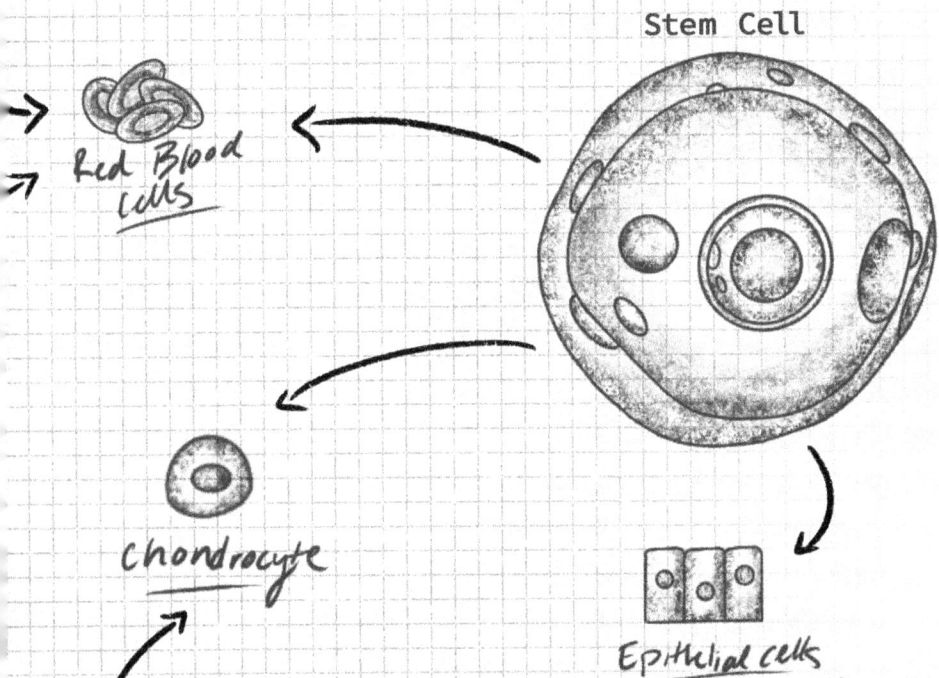

Stem Cell

Red Blood Cells

Chondrocyte

Epithelial cells

Dr. Austin Mardon, Sheher-Bano Ahmed, Samira Sunderji,
Olivia Brodowski, Dhwani Bhadresa, Pareesa Ali,
Si Cong (Sam) Zhang, Salma Abrahim, Rishi Thangarajah,
Anusha Mappanasingam, Rodala Aranya, Joonsoo Sean Lyeo

2021

First Printing: 2021

Typeset and Cover Design by Anna Kraemer

ISBN 978-1-77369-256-2

Golden Meteorite Press
103 11919 82 St NW
Edmonton, AB T5B 2W3
www.goldenmeteoritepress.com

Contents

Introduction

THIS BOOK WAS CREATED through the Antarctic Institute of Canada as a project sponsored by the Government of Canada's innovative Work-Integrated Learning program, Level Up. The Antarctic Institute of Canada is a non-profit Canadian charity organization founded in 1985 by former Antarctic researcher Austin Mardon. Its original aim was to lobby for the federal government of Canada to increase the extent of Canadian research in the Antarctic. Today, its objectives also include supporting scholarly research and academic writing.

A group of twelve postsecondary students worked on this book over a period of seven days. Each chapter was written by a different student, with some chapters being created through the collaborative efforts of multiple authors. All editing, graphic design, and audiobook production was also carried out by postsecondary students.

Thank you for taking the time to read this book and to learn a bit more about stem cells.

1

What Background Information Was Needed for the Discovery of Stem Cells?

*Written By **Samira Sunderji***

INTRODUCTION

Stem cells and stem cell research have gained popularity over recent decades due to their self-renewable properties and their ability to differentiate into other specialized cell types (Liu et al., 2020). However, the early stages of stem cell discovery date back to the 1950's, when this research revolutionized contemporary forms of medicine. Stem cell research has been a powerful force that has cultivated a reputation as one of the top medical achievements of the 20th century. In order to deepen our understanding of these cells, stem cell research continues to amaze the world as modern-day scientists and researchers innovate new technologies and ideas that continue to pave a pathway for successes and new discoveries. This chapter will divulge into the foundational aspects of stem cells and highlight the importance of stem cell research, with a particular focus on technologies and technological advancements that were used in the past and continue to be used today.

STEM CELL FOUNDATIONS

Stem cells, in the most basic of terms, are unspecialized cells that have the unique ability to transform and differentiate into specialized cells. Stem cells are able to self-renew (make replicas of the original stem cell) or differentiate into specialized cells that have a specific function including (but not limited to) muscle cells, skin cells, and brain cells. These cells are vital for the growth and repair of damaged organs or bodily tissues. Stem cells can be divided into 3 categories consisting of embryonic stem cells, induced pluripotent stem cells, and tissue stem cells. These categories and more information on the specifics of stem cells will be thoroughly discussed in chapter 5.

STEM CELL RESEARCH HIGHLIGHTS

Stem cell research offers great promise for understanding the basic mechanisms behind human development, as well as hope for new treatments and therapies. Debilitating diseases and disorders that were once thought to be incurable such as Parkinson's disease, have inspired scientists and researchers worldwide to continue on the road of development and innovation.

News and media reports fuel the public's interest in this new type of regenerative medicine however, entertwined in the media frenzy and the increasing public interest of stem cell research are numerous political agendas and various religious and ethical concerns (Bongso & Richards, 2004). For example, human embryonic stem cell (hESC) research is ethically and politically controversial as it involves the destruction of human embryos. To further fuel this debate, embryonic stem cell research is often unjustly grouped into the realm of reproductive cloning, which is an entirely different topic altogether (Bongso & Richards, 2004). Further controversies and technical, ethical, and legislative issues will be discussed in chapter 10. Nonetheless, the historical success of stem cell research in regards to innovative and prospective treatments and therapies in reparative medicine has greatly impacted the lives of millions across the globe for the better, and there is good reason to be optimistic for the future (Bongso & Richards, 2004).

THE USE OF MICROSCOPY IN STEM CELL RESEARCH

The discovery and continued understanding behind the mechanisms of stem cells and the different purposes they serve through their various roles in the body is attributed to the hard work of scientists over the past few decades. However, the tools and technologies used to facilitate the discovery and studies of stem cells are equally as important to credit.

Microscopy involves the study of objects that are unable to be examined by the unaided eye. The microscope is an instrument which provides an enlarged image of the object that is being studied. The word microscope is derived from the Greek language–"mikos" in Greek is translated to "small", and "skopein" translates to "see" or "looking at". In the SI metric system of units, the sizes of objects using a microscope are expressed in terms of sub-multiples of the meter, including the micrometer (also referred to as a micron) and the nanometer (Masters, 2009).

The Electron Microscope (EM)

The electron microscope (EM) has been an invaluable tool in the field of biomedical research for over 75 years through its use in studying the fine properties and structures of cells and tissues
(Vemuri, 2007). While the origin story of the electron microscope is complex, the contributions of various researchers were required to turn the initial idea of microscopy into the workable instrument used in laboratories today (Robinson, 1986). It was discovered that electrons have a dual nature with the ability to behave as particles or waves. The shorter wavelength of electrons was the basis on which the electron microscope was invented, as a shorter wavelength allowed for the better resolution of smaller, minute details. This is why the optical microscope, invented in 1590 by Hans and Zacharias Janssen, had ceased to become useful beyond a particular magnification, as the optical waves began to pass around the lenses with poor resolution.

There are two types of electron microscopes; transmission electron microscopes (TEM) and scanning electron microscopes (SEM). Transmission electron microscopes are used to view thin specimens (such as the interior of cells and the structure of protein molecules) through which electrons can pass and this generates a projection image (UMASS,

2018). Scanning electron microscopes rely on projecting a focused beam comprised of high-energy electrons onto to the surface of a specimen, thereby providing detailed images of the surface of cells and whole organisms that are not possible by TEM (UMASS, 2018).

The Transmission Electron Microscope (TEM)

Under the guidance and mentorship of Max Knoll, physicist Ernest Ruska invented the transmission electron microscope for his Doctoral thesis at the Technical University of Berlin in 1931. This two-stage transmission electron microscope was constructed using magnetic lenses with a magnification of 16X and a wire mesh acted as the object under examination (Bogner et al., 2006). In 1933, Ruska improved upon his original design model and demonstrated its full capabilities to surpass the resolving power of the light microscope (Masters, 2009). He built a transmission electron microscope with three magnetic lenses–one of which acted as a condenser to gather the electrons and focus them onto the object, illuminating the area being examined, while the other two lenses, the objective lens and the projection lens, were used to focus on the object and magnify the object onto the viewing screen, respectively (Masters, 2009; UMASS, 2018). A camera located outside the microscope had the ability to photograph the magnified image on a viewing screen (Masters, 2009). The microscope's total magnification was 12 000 X and the rotating object stage could hold up to 8 specimens (UMASS, 2018). Even though Ruska's work began in the 1930's, it was only in 1986 when Ruska was awarded the Nobel Prize in Physics for the invention of the transmission electron microscope (Masters, 2009). Nonetheless, his efforts advanced the world of science and contributed to the success of his predecessors who continue his research and innovative legacy.

The Scanning Electron Microscope (SEM)

On the other hand, the invention of the scanning electron microscope cannot be pinpointed to only one contributor or researcher. The earliest known work describing the concept of a SEM was by Max Knoll, who built a "scanning microscope" in 1935 for the study of the targets of television camera tubes. However, without the use of demagnifying lenses, there was a limit on the resolution and quality of the resulting image (Bogner et al., 2006). The first true SEM was invented in 1942 by Vladmir Zworykin, a Russian-American inventor and engineer. The microscope was used to examine the surface of a solid specimen in great detail (Bogner et al., 2006). He demonstrated the use of secondary electrons which provided topographic contrasts to images, and introduced the use of an electron multiplier tube which is still used today as vacuum electronic devices, also known as photomultipliers (PMTs) (Bogner et al., 2006). Zworykin's improvements provided a framework for success and this allowed for the creation of various prototypes and improvements similar to the SEMs that are used in research today.

Using Electron Microscopes for Stem Cell Research and Discoveries

In the context of stem cells, high-resolution imaging by transmission electron microscopy has proven to be extremely useful in studying the fine, structural details of cells and cellular organelles (Miko et al., 2015). A 2015 study on mesenchymal stem cells derived from bone marrow (that function to create and repair skeletal tissues) used TEM to obtain detailed morphological descriptions of different adult tissue-derived stem cells. It was found that the TEM, along with the researchers' histological approach to process cells, was very efficient in producing high quality images of cell populations (Miko et al., 2015). While the preparation process for this form of microscopy is beyond the scope of this chapter, TEM is an essential and valuable tool to deepen the understanding of adult stem cells derived from different populations (Miko et al., 2015). This information is especially important in regenerative medicine and tissue engineering. Scanning electron microscopy, with its three-dimensional (3D) perspective, magnification, and depth of field, can procure vivid and readily interpretable images (Vemuri, 2007). The combination of a high-resolution SEM and specific techniques during sample preparation of stem cells has the ability to reveal detailed biological surfaces of said cells (Vemuri, 2007). Another application of electron microscopes can be seen through their role in the discovery of microvesicles released from human embryonic stem cell derived-mesenchymal stem cells (hESC-MSCs). This type of stem cell is able to inhibit the proliferation of cancer-causing leukemia cells. Microvesicles are released from these cells and play an important role in antitumor activity (Ji et al., 2017). Through the use of SEMs and TEMs, it was found that microvesicles released from hESC-MSCs inhibited the growth of tumors and stimulated autophagy, a process in which the human body disposes of damaged cells to regenerate newer and healthier cells (Ji et al., 2017). The paper alluded to the fact that excessive autophagy might induce apoptosis, or cell death, of the cancer-causing leukemia cells (Ji et al., 2017). Thus, the use of electron microscopes in the context of stem cells have allowed for new findings to be reported in literature, which proves its usefulness in modern-day applications of scientific research.

THE USE OF GENOME EDITING IN STEM CELL RESEARCH

Genome editing, also referred to as gene editing, is extensively used in the field of biological and biomedical research. Focusing on stem cell research of human diseases, this application of genome editing can allow for the development of novel therapeutic agents as treatments and therapies (Lee et al., 2020). Genome editing allows for the alteration of an organism's DNA. Generic genome editing strategies involve DNA modifications in live organisms which include deletions of DNA, corrections via gene replacement, and insertions of DNA (Lee et al., 2020). Currently, there are two main genome editing approaches used in laboratory research; ex vivo and in vivo genome editing. Ex vivo genome editing refers to the process of genetically editing a cell external to the living organism, which is then transplanted back into the organism following the editing process. In vivo genome editing involves a more direct approach in which the gene therapy is administered intravenously or locally to a specific organ of interest (Ex Vivo & In Vivo Gene Therapy Techniques, 2020). The ex vivo approach is used more frequently than the in vivo approach due to advantages in editing, safety, and overall higher efficiency (Lee et al., 2020).

Large-scale genome editing technologies have the ability to link genetic variants with diseases and this proves to be useful in real-world applications of genetic testing and screening for at-risk individuals. Some technologies include zinc finger nucleases (ZFNs), transcription activator-like effector nucleases (TALENs), and clustered interspaced short palindrome repeats, commonly referred to as CRISPR (Lee et al., 2020). Despite the unique characteristics of each tool, they share a common underlying principle: they all separate chromosomal DNA in site-specific areas, which leads to more precise editing (Lee et al., 2020). This approach to genome editing with engineered nucleases was named the 2011 "Method of the Year" by Nature *Methods*, a peer-reviewed science methodology journal that publishes novel laboratory techniques and methods in the field of life sciences (Joung et al., 2013).

Zinc Finger Nucleases (ZFNs)

ZFNs are a powerful tool used for site-specific alterations and precise modifications of the human genome (Ramalingam et al., 2013). ZFNs were one of the first programmable nucleases (enzymes that produce site-specific DNA double-stranded breaks) to be used in the process of stem cell genome editing (Lee et al., 2020). The vast majority of literature that describes the process of genome editing has been performed using ZFNs (Joung et al., 2013). ZFNs are composed of two domains, both of which have the ability to bind DNA sequences of different base pairs in length (Lee et al., 2020). The first is a DNA-binding zinc-finger domain that binds up to 18 DNA sequence base pairs and is fused to a bacterial nuclease (Lee et al., 2020; Ex Vivo & In Vivo Gene Therapy Techniques, 2020).The second domain is a nuclease domain derived from the FokI restriction enzyme. This restriction enzyme recognizes a specific DNA sequence and cleaves the genome at that location. A pair of ZFNs are used together to target and cleave (cut) the specified DNA (Ex Vivo & In Vivo Gene Therapy Techniques, 2020). ZFNs are able to alter genomes via point mutations, deletions, insertions, inversions, duplications, and translocation of a DNA sequence (Joung et al., 2013). However, there are some drawbacks to this technique–laboratory procedures using ZFNs are time consuming and require a significant amount of effort to produce an effective edit (Ex Vivo & In Vivo Gene Therapy Techniques, 2020).

Transcription Activator-Like Effector Nucleases (TALENs)

TALENs are similar to ZFNs in regard to their overall structure. They are derived from naturally occurring plant pathogenic bacteria and contain DNA binding proteins. These binding proteins, formally known as TALEs, are composed of 30-35 amino acids and have the ability to recognize a single base pair of DNA. Similar to ZFNs, multiple TALEs form a chain when linked together and are capable of targeting a single site within a particular genome. Another similarity between ZFNs and TALENs is its use of FokI as the nuclease, which produces a double stranded break in double-stranded DNA. TALENs were first used for genome editing of bacterial proteins. However, in principle, TALENs can edit any type of genome sequence (Lee et al., 2020).

Clustered Regularly Interspaced Short Palindromic Repeats (CRISPR)

The CRISPR approach was identified in 1987 in Escherichia coli (E. coli) by researcher Yoshizumi Ishino and his colleagues in 1987 in Osaka. However, the application to genome editing was demonstrated for the first time in 2013 (Lee et al., 2020). This technique has evolved rapidly over the last few years. The most common CRISPR systems that are currently used are divided into two major classes and encompass six types. The CRISPR/Cas9 system is the most documented CRISPR system as it was repurposed for site-specific genome manipulation rather than a system of adaptive immunity against the *E.coli* invading pathogen.

Using Genome Editing for Stem Cell Research and Discoveries

The CRISPR/Cas 9 system has been applied to stem cells via the genetic modification of pluripotent cells (that have the ability to differentiate into any cell) and multipotent cells (that have the ability to only develop into some specific cells) (Patmanathan et al., 2018). These cells, once differentiated, are used for functional analysis and clinical transplantation. (Patmanathan., 2018). For example, a 2019 study found that the CRISPR/Cas 9-based system for genome editing played a significant role in generating human skin equivalents (HSEs) via pluripotent stem cells and other tissue engineering techniques (Jacków et al., 2019). This provides a glimmer of hope for those diagnosed with recessive dystrophic epidermolysis bullosa (RDEB), a severe inherited skin disorder caused by mutations in the gene that codes for type VII collagen, a type of collagen necessary for strengthening and stabilizing the skin (Jacków et al., 2019). While this clinical trial used mice, the results from this study serves as a foundation to translate this treatment into a viable option for humans diagnosed with RDEB (Jacków et al., 2019). Another application of genome editing in stem cell research can be seen through its role in studying allogeneic human umbilical mesenchymal stem cells (alloMSC) as a cell source to treat heart attacks (Jacków et al., 2019). The 2018 study demonstrated that human alloMSC (stem cells derived from the mesoderm that are transplanted from one individual to another) with beta-2 microglobulin (B2M) knockdown (a marker for cancerous tumors ad adverse cardiac events) by CRISPR/Cas 9 gene editing is a convenient and useful cell source in treating heart attacks without inducing immune response rejections by the body (Jacków et al., 2019).

CONCLUSION

Stem cells have proven their worth as a viable treatment option in the fight against various degenerative and life-threatening diseases. With the use of in-depth stem cell research and technology, unique insights into the cellular mechanisms of stem cells can be further understood and can be used in various diagnostic and therapeutic applications. While the credit for these successes can be attributed to world-renowned scientists and researchers, numerous technologies and tools that have been created in the last century are also important to acknowledge. From the electron microscope to genome editing tools, these technologies have facilitated remarkable improvements to contemporary forms of equipment. What were once simple, unrealistic ideas, have morphed into groundbreaking medical achievements, leading to a plethora of literature and experimental studies on stem cells. The next chapter will focus on the discovery of stem cells, with a particular focus on the key milestones of stem cell research over the last few decades.

2

How Were Stem Cells Discovered?

Written By **Olivia Brodowski**

INTRODUCTION

Stem cells and stem cell research has become more popularly featured in media and scientific discourse in the last few decades. However, the history and discovery of stem cells spans multiple decades. This chapter will explore the discovery and research of stem cells and related topics dating back to the 1950s all the way to the 21st century.

THE DISCOVERY OF STEM CELLS

While the idea of 'stem cells' was proposed by embryologists in the early 19th century, it was not until the mid-19th century that two Canadian scientists: James Edgar Till and Ernest Armstrong McCulloch "collaborated on a series of experiments in which they injected bone marrow cells into heavily irradiated mice" (Steesman & Kyle, 2021, p. 830). After the marrow cells were injected, the two researchers noticed that "small nodules grew on the spleens of the irradiated mice" (Steesman & Kyle, 2021, p. 830). These nodules were named spleen colonies. Till and McCulloch correlated these nodules with the injection of the marrow cells because the number of nodules closely matched the number of marrow cells injected. Further observation led to them obtaining evidence "that some (but not all) of the spleen colony cells were also self-renewing... [a] critical property of a stem cell" (Steesman & Kyle,

2021, p. 830). For the groundbreaking discovery of stem cells, Till and McCulloch were both granted a title in the Canadian Medical Hall of Fame in 2004.

THE 1950S

Formal stem cell research began in the early 1950s. The study of teratocarcinomas, embryonal carcinoma cells, and embryonic stem cells is some of the earliest research done regarding stem cells, although scientists at the time were unaware of it. Teratocarcinoma embryonal carcinoma cells are a form of cancerous, germ cell tumor that occurs in the testicles of both animals and humans. Embryonic stem cells are cells from the undifferentiated inner mass cells of a human embryo.

What is a teratocarcinoma cell? The Greek word, 'teratos' (monster), describes the appearance of teratocarcinomas cells, "as these tumours are composed of haphazard mixture of adult tissue and misshapen organs… containing teeth, pieces of bone, muscles, skin and hair" (Solter, 2006, p. 319) in some cases. Researchers, Leroy Stevens and Barry Pierce, "assembled a detailed picture of the biology of spontaneous testicular teratocarcinoma" (Solter, 2006, p. 319). This detailed picture showcased how Stevens followed a small group of abnormal cells mutate, develop, and grow into cancerous ones.

Moreover, Stevens and Pierce identified that "a single cell derived from a tumour and injected intraperitoneally [directly into the body] can produce all cell types that are encountered in a teratocarcinoma" (Solter, 2006, p. 320). The discovery that teratocarcinoma cells have the ability to produce other cell types is a fundamental early discovery in stem cell research because it demonstrated that teratocarcinomas possessed a specific type of stem cell, "a single one which has the capacity to grow indefinitely [and]… differentiate into multiple adult cell types" (Solter, 2006, p. 320). The pluripotency of teratocarcinoma cells and the observation of features of "the tumours that are called embryoid bodies" (Solter, 2006, p. 320) indicated that teratocarcinomas originated from embryonic cells. Further observation confirmed that "pluripotent stem cells of early mouse embryos and teratocarcinomas might be highly similar, if not identical" (Solter, 2006, p. 320), introducing the early concept of embryonic stem cells.

Throughout the 1950s and 60s researchers hyperfocused on understanding the relationship between teratocarcinoma and embryonic stem cells. Primarily, their focus was in triggering and observing teratocarcinomas and extracting, cloning, and implantation of teratocarcinomas into other mice to observe patterns of differentiation. Little work has been similarly carried out on human teratocarcinoma tumours.

Further pivotal research that occurred in the 1950s includes Strickland and Maldavi's discovery that chemical compounds "can induce the nullipotent embryonal carcinoma cell... to differentiate" (Solter, 2006, p. 321). Nullipotent cells do not go through cell division and, thus, are terminally differentiated. Strickland and Maldavi's discovery was one of the first moments where scientists chemically triggered differentiation in cells. Strickland and Maldavi's findings were groundbreaking because they introduced the possibility of specialized induced stem cell differentiation. For the first time scientists were met with the possibility of histological-level control over mutation and cellular modification.

EARLY STEM CELL THERAPY

In the early 1950s scientists proposed "that a stem cell may be connected to the origin and evolution of myeloid leukemia" (Jordan, 2007, p. 13) Leukemia is a cancer that forms in the blood-making tissues such as the lymphatic system and bone marrow. Scientists hypothesized that leukemia develops through the mutation of a normal stem or progenitor cell which then "giv[es] rise to an entity that is functionally defined as a leukemia stem cell" (Jordan, 2007, p. 13). Progenitor cells are like normal stem cells because they undergo differentiation but are different because they are more specified, already having a 'target' tissue to which they are differentiating. While the normal stem cell continues to undergo its proper differentiation, the mutated stem cells share properties with normal stem cells. The mutated or leukemic stem cell "can undergo self renewal, are multipotent, and highly proliferative" (Jordan, 2007, p. 13). Mutations can occur all throughout the cell differentiation process, both in normal stem cells and progenitor cells that no longer have stem cell characteristics. This shows that leukemic stem cells which can maintain the leukemic disease can mutate spontaneously. This spontaneity is one reason why leukemic disease is difficult to treat in the human body and also why it is so difficult to observe and study leukemic cells: they are constantly changing.

In 1957 E. Donnal Thomas was the first to publish his findings on radiation and chemotherapy as well as the intravenous infusion of bone marrow. Intrigued with the discovery that lethally irradiated mice were able to survive as a result of marrow injections, as previously discussed, Thomas "become convinced of the clinical potential of human marrow transplantation" (Appelbaum, 2007, p. 1472). Thomas's initial experiences with allogeneic marrow transplantation in humans was unsuccessful, with patients only surviving an additional 100 days—albeit with no additional adverse effects. Research regarding histocompatibility was underdeveloped and so patient and donor matching was not considered, surely resulting in the relative unsuccess of Thomas's trials. Thomas was not discouraged and continued his research and found that proper donor selection was key after "conducting experiments using an outbred canine model" (Appelbaum, 2007, p. 1472). His findings after these experiments concluded "that most dogs given sufficient irradiation followed by grafts from dog *leukocyte antigen-matched littermates* and a short course of immunosuppression survived long-term" (Appelbaum, 2007, p. 1472 emphasis added). This was a key discovery because it supported the importance of donor matching in future stem cell research regarding rehabilitative procedures.

After this, Thomas returned to human marrow grafting which concluded with him and his team successfully demonstrating the survival "of patients who received transplants for end-stage acute leukemia" (Appelbaum, 2007, p. 1473). The success of being able to cure a small percentage of previously incurable patients encouraged him and his colleagues to explore transplant procedures at earlier stages of the disease, "and in 1979 he and his colleagues reported achieving a cure rate above 50%" (Appelbaum, 2007, p. 1473). Initially, marrow transplantation was limited to donors who were matching siblings or identical twins; however, "[i]n the late 1970s, the Seattle group performed the first successful marrow grafting from a matched, unrelated donor in a patient with leukemia, which helped to stimulate the formation for the National marrow Donor Program and the subsequent registration of more than 11 million volunteer donors" (Appelbaum, 2007, p. 1473).

THE 1970S AND 80S

Ralph L. Brinster is an American geneticist whose research in embryonic stem cell differentiation revolutionized mechanisms of gene control and stem cell physiology. In 1974 Brinster successfully tested that "embryonal carcinoma cells retain their embryonic nature" (Solter, 2006, p. 322) when injected directly into a mouse's blastocyst cavity. Further examination of Brinster's discovery found a discrepancy in the development of embryonal carcinoma cells. Some cells would develop without the appearance of abnormal tumours, while others "reported limited chimerism, recurrent development of tumours derived from embryonal carcinoma cells" (Solter, 2006, p. 323). In 1981 the first mouse embryonic stem cells were "derived directly from the mouse blastocyst." (Solter, 2006, p. 320). These embryonic stem cells differentiated less sporadically and did not undergo the same spontaneous mutation that the embryonal carcinoma cells exhibited. Once embryonic stem cells were successfully isolated, the issue regarding abnormal development of embryonal carcinoma cells became irrelevant. At this time, research and focus on embryonal carcinoma cells became less and less significant.

KEY MILESTONE: 1989

In 1989 "[t]he first genetically modified mouse [was] generated using homologous recombination in embryonic stem cells" (Solter, 2006, p. 321). This is a key milestone in the discovery and research of stem cells because it represents scientists' ability to maintain a high degree of stability over— what was previously thought as spontaneous— cell differentiation.

EMBRYONIC STEM CELLS

In 1969 Dr Martin Evans, a British Biologist, began working with a group of Leroy Stevens' mice, some of which were carriers of the transplantable teratocarcinomas. Evans observed that the tissues that made up the transplanted teratocarcinomas were so well differentiated that it was, in fact, difficult to find the stem cell of the tumour. As a result of this observation, Evans was able to isolate "clones of small, piling, rapidly growing cells, from which [he] was able to establish mass cultures" (Evans, 2011, p. 681). Alongside Matt Kaufman and Gail Martin, Evans "found that cells from early mouse embryos exposed to the same culture environment can suspend developmental progression and continue to multiply while remaining pluripotent" (Smith, 2010, p. 113), thus discovering embryonic stem cells.

Soon after the isolation of mouse embryonic stem cells, researchers began studying the ability of mouse embryonic stem cells to form germline chimaeras. Germline chimerism happens when the germ cells of an organism— like the sperm or egg— are not exactly identical to the cells of the original organism. For example, a chimera would exist in a mouse with a teratocarcinoma implantation because the teratocarcinoma cells do not come from the original organism— the mouse. A human could be a chimera, for example, after having a bone marrow transplant because the marrow cells do not stem from the original organism and the subsequent red blood cells that are produced from that marrow are from a different originating cell. This study concluded that mouse embryonic stem cells "are extremely efficient in contributing to germline chimaeras" (Solter, 2006, p. 323-324). As a result, researchers have been able to develop "several thousands of mutations and can expect that a mutation for every mouse gene will soon be available" (Solter, 2006, p. 324).

Recent interests in the research of embryonic stem cells as a possible solution to "developing cell and tissue therapies in humans prompted a renewed interest in mouse embryonic stem cells, how to culture them, and how to control their in vitro differentiation" (Solter, 2006, p. 324). The standard culture— the nurturing environment— for embryonic stem cells includes the soluble factor: leukaemia inhibitory factor, which helps prevent the embryonic stem cells from spontaneously differnitatieing, which they are prone to do in vitro. Conducting such experiments on human embryonic stem cells is not as easily accessible.

KEY MILESTONE: 1998
In 1998 scientists conducted the first derivation of "human [embryonic stem] and embryonic germ cell lines" (Solter, 2006, p. 321).

HUMAN EMBRYONIC STEM CELLS
 Despite thoroughly developed research practices and an understanding of mouse embryonic stem cells, research regarding human embryonic stem cells took longer to develop most likely due to "the difficulties involved in obtaining suitable human embryonic material and an under-standable reluctance of most investigators to work in a field that is fraught with potential legal problems and political and moral dilemmas" (Solter, 2006, p. 324).

2006— INDUCED PLURIPOTENT STEM CELLS
Dr Shinya Yamanaka was the first to demonstrate "that differentiated somatic cells can be de-differentiated into induced pluripotent stem cells" (Li et al., 2013, p. 3594). This is a fundamental discovery in stem cell research because it "raise[s] the prospect of better patient care by regenerating virtually all types of somatic cells from patient's own tissue"

(Li et al., 2013, p. 3594). In 2009, Yamanaka introduced a state-of-the-art method for reprogramming somatic cells because knowledge regarding the exact molecular mechanisms of somatic cells was underdeveloped. The purpose of reprogramming somatic cells is to be able to control differentiation in already differentiated cells. This would resolve the necessity of acquiring embryonic stem cells to perform differentiation. Instead, virtually all types of cells would be able to differentiate, making the research process easier as well as making stem cell therapies more available. This method of reprogramming is called the elite stochastic model. The previous model— the elite model— "predicts that only certain cells are possibly reprogrammed, whereas the stochastic model predicts that most or all somatic cells have the potential to become pluripotent" (Li et al., 2013, p. 3594).

At its core, the stochastic model suggests the importance of epigenetic modification and anti-senescence in induced pluripotent stem cells. Epigenetic "changes are reversible and do not change your DNA sequence, but they can change how your body reads DNA sequence" (CDC, 2020). Senescence is when a cell stops dividing at maturation but does not die. Anti-senescence, then, is the process of the continuation of cell division even at full maturation. The stochastic model explains that "reprogramming stimuli must push the cells up into the pluripotent state by changing epigenetic status, and this is achieved by locking the induced pluripotent stem cell in the pluripotent zone by specific epigenetic modification" (Li et al., 2013, p. 3594). Scientists have gathered evidence that shows that the various stimuli involved in this process include the "DNA methylation, histone acetylation, and histone methylation" (Li et al., 2013, p. 3595).

Moreover, the stochastic model's insight regarding epigenetic transformation has prompted scientists to research epigenetic memory. The concept of epigenetic memory rises from the necessary eraser of the original epigenetic signature of the somatic cell during the reprogramming process. Epigenetic memory introduces "the notion that erasing

[the] epigenetic pattern of donor cells" (Li et al., 2013, p. 3596) is vital for induced pluripotent stem cell generation.

CONCLUSION

In conclusion, the discovery of stem cells has been a venture of meticulous experimentation. Researchers and scientists have built off of each others work for the past half century, incrementally gathering information and evidence regarding the potential benefits of stem cell research on human health and therapy options. Starting with meticulous experimentation on mice and now progressing towards human therapeutic research, stem cell research has gone a long way but, evidently, has still a long way to go. While progress in the field of stem cell research is at a slow pace due to the lack of public funding and other contentions, the effects and implication of stem cell research on society are not few and far between.

3

What is the Impact of Stem Cells on Society?

*Written By **Dhwani Bhadresa***

S TEM CELLS, THE PRODUCT OF great amounts of scientific research and analysis, have become indispensable tools in modern medicine. Stem cells have vast applications in both the medical and biological sciences and could possibly become "one of the most important aspects of medicine" in the future (Zakrzewski et al., 2019). This chapter will focus on understanding the complexities of stem cells and their impact on society. In order to comprehend the influence of stem cells, it is essential to recognize how society influences medicine. Specifically, we will discuss the importance of stem cells' influence on regenerative medicine, bone marrow transplants, dentistry and neurodegenerative diseases.

Medicine requires social factors, context and circumstance to be practiced and applied to society (Amzat & Razum, 2014). Society defines what kind of medicine is practised and how it is shaped within individuals (Little, 2016). Institutional structures and societal factors will impact medical conditions in terms of the delivery of care and organisation (Amzat & Razum, 2014). Thus, it is important to understand that society and medicine work together, to benefit individuals within the collective. Stem cells are an innovative discovery that allow society to better understand disease and potentially find a cure that will ultimately impact policy, social structure, and health care.

Stem cells' impact on society can be analyzed through the influence on regenerative medicine and clinical practices (Kolios & Moodley, 2013). Regenerative medicine stimulates, replaces, and regenerates cells or organs using stem cell therapy (Kolios & Moodley, 2013). This regeneration can only take place due to the capabilities of stem cells to differentiate into various types of cells and develop into any form of tissue within the body (Biehl & Russell, 2009). This has an enormous impact on society due to the applicability of stem cells on therapeutic treatments in relation to repairing and regenerating the tissue (Biehl & Russell, 2009). This ability opens up a broad range of possible stem cell-based therapies that could have the potential for radically transforming the ways in which modern medicine is conceptualized and practiced, as seen with tissue regeneration therapy. One of the possible therapies that exist is organogenesis which is the process in which an organ is developed through the embryo tissue layers mesoderm, ectoderm and endoderm (Kolios & Moodley, 2013). Stem cell based organogenesis can combat the lack of organ donors through stem cell generated organs, which allows for greater availability for organ transplants.

There are specific types of stem cells that can be used for regenerative purposes including embryonic stem cells and induced pluripotent stem cells.

Embryonic stem cells specifically can be utilized as an exceptional method to further comprehend organogenesis and human development (Kolios & Moodley, 2013). Induced pluripotent stem cells have the ability to develop into any type of cell within the human body and will be important in studying safe and novel therapies for various diseases (Kolios & Moodley, 2013). In terms of societal impact, this means that stem cells may have the capability and capacity for organ regeneration and replacing tissues that have been injured (Kolios & Moodley, 2013).

Additionally, induced pluripotent stem cells aid the medical community in improving the analysis of genetic causes of conditions, known as the pathogenetics of disease (Kolios & Moodley, 2013). This is through the creation of human models of conditions through utilizing induced pluripotent stem cells (Kolios & Moodley, 2013). This modeling and improvement in pathogenetic knowledge facilitates further advances for the treatment of degenerative conditions via cell based therapy (Kolios & Moodley, 2013).

As with any form of medicine there are limitations in regard to regenerative medicine and cell therapy (Kolios & Moodley, 2013). Through stem cell treatment, immune rejection can occur, when the body is attempting to protect itself via attacking the body's stem cells (Kolios & Moodley, 2013). In order to avoid immune rejection, mesenchymal stem cells, induced pluripotent stem cells, and tissue from the placenta have aided in circumventing this type of immune response (Kolios & Moodley, 2013). In addition, there is lack of information on stem cells and genetic stability, which is precise DNA replication and repair systems of cells (Bhagavan & Ha, 2015, p. 22) specifically in the case of induced pluripotent stem cells (Kolios & Moodley, 2013). In cases of genetic instability this can lead to the development of tumors (Kolios & Moodley, 2013). Stem cells characteristics including self-renewal and plasticity (stem cells capability to differentiate into other cell types) can potentially correlate to the development of cancer also known as carcinogenesis within the host cells or tissue (Kolios & Moodley, 2013). Through therapies such as cell transplantations, with the use of embryonic stem cells and induced pluripotent stem cells the possibility of developing a tumor is present (Kolios & Moodley, 2013). It is important to keep in mind that stem cells can have negative side effects impacting the individual subjects who are undergoing this form of treatment.

One of the most common uses of stem cells is through bone marrow transplants, also classified as hematopoietic stem cell transplant (Khaddour et al., 2021). A hematopoietic stem cell can be located in the bone marrow and the periphery of the blood (Definition of Hematopoietic Stem Cell—NCI Dictionary of Cancer Terms—National Cancer Institute, 2011). Additionally, these cells have the capability to differentiate into all forms of blood cells, specifically platelets, red blood cells, and white blood cells (Definition of Hematopoietic Stem Cell—NCI Dictionary of Cancer Terms—National Cancer Institute, 2011). Patients with defective or dysfunctional bone marrow, undergo this process of transplantation for the purpose of acquiring healthy hematopoietic stem cells (Khaddour et al., 2021). The benefits of this transplant directly impacts populations diagnosed with the following: tumors, immune deficiency disorder, leukemia, lymphomas, myeloma, as well as other bone marrow diseases (Bone Marrow Transplantation, n.d.). Through the transplantation of stem cells, patients have the possibility for the disease to be cured as healthy stem cells replace non functioning cells, and restimulate the immune system (Bone Marrow Transplantation, n.d.). Furthermore, if the transplantation is successful this can aid in increasing functionality of the existing bone marrow and could result in tumor cell destruction or functional cell production dependent on the condition (Khaddour et al., 2021). This transplant impacts societies populations that have been impacted by cancer and diseases (Bone Marrow Transplantation, n.d.). Cancer predominantly impacts the older populations, thus finding a means of more effective treatment allowing for a greater quality of care. This can increase survival rates and improves patient outcomes.

Societal need for hematopoietic stem cell transplantations has substantially increased over the years. As is evidenced by the increase in the number of hematopoietic stem cell transplantations that have taken place over January 1, 2006 and December 31, 2014 (Gratwohl et al., 2015). The Worldwide Network for Blood and Marrow Transplantation assembled large amounts of data on the activity of transplantations (Gratwohl et al., 2015). In over 75 countries, 953 651 haemopoietic stem cell transplantations took place across 1516 transplant units (Gratwohl et al., 2015). During data collection, it was noted that transplants were not reported in countries with a surface area smaller than 700 km2, a population of 300 000 or lower and if the income per person was less than $1260 USD (Gratwohl et al., 2015). Additionally, by 2012, unrelated donor registries were making a huge impact when approximately 22.3 million volunteers were donors, and the amount of core blood products was estimated at 645 646 (Gratwohl et al., 2015). Unrelated donor registries are registries of donors who are unrelated to the recipient but are still eligible to donate. Through the advancement in stem cell technology, our society was able to perform and report almost one million transplants over eight years. With this technology, countless individuals and societies across the world were able to receive adequate treatment for their diagnosis. Were able to come together to donate through acts of altruism to aid societal members.

The first transplant of hematopoietic stem cells took place in 1957, and by 1985 the number of transplants had increased to approximately 10 000 (Gratwohl et al., 2015). That number had increased to 100 000 transplants by 1995 and an estimated one million transplants in December of 2012 (Gratwohl et al., 2015). There are certain factors associated with increase in transplants specifically, rates increased in countries with accessibility to resources, the presence of additional transplant teams and structures set up for unrelated donors (Gratwohl et al., 2015). However, unequal access to medical care due to a lack of resources negatively impacts the health outcomes of individuals and societies. For example, among other things, a lack of qualified professionals, research facilities and medical technology precludes many in developing countries from being able to partake in the access to stem cell research. Currently ongoing clinical trials of hematopoietic stem cell transplants are trying to assess the efficacy of using stem cells to treat various medical disorders (Khaddour et al., 2021). The more information gathered, and trials conducted, the more opportunities for stem cells to aid societal populations that are impacted by life threatening disease to provide an avenue of treatment.

Stem cells have aided in treating medical conditions specifically for marginalized people including Hispanic and African American populations (Khaddour et al., 2021). It also benefits individuals in countries where there is a lack of accessibility to medical resources (Khaddour et al., 2021). The specific treatment in reference is haploidentical stem cell transplantation, which facilitates transplant of bone marrow product into the recipient from the donor (Khaddour et al., 2021). This differs from other bone marrow transplants because the donor is only qualified as a half match meaning the donor relative is only 50 % genetically related to the patient (Khaddour et al., 2021). This is classified as a haplotype-mismatched donor (Khaddour et al., 2021). The advantages of this treatment is reducing costs and quick accessibility and availability of specific hematopoietic cell products (Khaddour et al., 2021). There are certain disadvantages to this form of action such as complications like Graft versus host disease, which in turn leads to higher rejections of grafts and can lead to mortality (Khaddour et al., 2021).

It is critical to understand how the administration of this specific type of transplantation focuses on aiding individuals of colour in society in reference to their medical conditions. In the US, there are approximately 9000 new diagnoses of chronic myelogenous leukemia and the best form of treatment is bone marrow transplants or stem cells ("Racial Minorities Face a Dearth of Stem Cell Donors," 2019). In order for these treatments to work a donor is required that can be matched to the patient. This process of procuring a donor for racial minorities is decreased in the United States in comparison to white americans, because less non-white individuals are registered as donors ("Racial Minorities Face a Dearth of Stem Cell Donors," 2019). Thus, it is important to emphasize forms of treatments that can aid individuals of colour and highlight the significance of becoming a donor.

Stem cells also have a role in the field of dentistry as regenerative medicines. (Zakrzewski et al., 2019). Teeth, in particular, serve as advantageous for the procurement of stem cells due to their capacity to function as a non-invasive and natural source of stem cells (Zakrzewski et al., 2019). Dental tissue that is functional and healthy contains large volumes of stem cells (Zakrzewski et al., 2019). In particular, there are three stem cells that demonstrate profound impacts on the medical community: dental pulp stem cells, periodontal ligament stem cells and stem cells from the apical papilla (Zakrzewski et al., 2019).

The location of dental pulp stem cells are isolated in dental pulp which is the most inner layer of the teeth (Zakrzewski et al., 2019). These stem cells can facilitate chondrogenic and osteogenic potential (Zakrzewski et al., 2019). Chondrogenesis is the ability to process the development of cartilage and osteogenic is the formation of bone. The application of these stem cells can be possibly utilized in a variety of ranges, specifically reconstruction relating to orthopedics, jaw, face, and even outside the confinements of the oral cavity (Zakrzewski et al., 2019). In addition, dental pulp stem cells have the capability to "generate all structures of developed" teeth (Zakrzewski et al., 2019). Furthermore, these stem cells can aid in treating conditions that have neurological deficits, because under the right circumstances, these cells can turn into neural cells (Zakrzewski et al., 2019). These structures are necessary for the function of oral care and thus, when teeth become damaged, stem cells have the possibility to regenerate and reconstruct the damage from the disease aiding in the health of societal members.

The location of periodontal ligament stem cells is within the periodontal ligament, these cells can be found on the outside surface of alveolar bone and the root of the tooth (Zakrzewski et al., 2019). These stem cells have already been deemed useful for treatment of periodontal regeneration therapy due to the efficiency and safety (Zakrzewski et al., 2019). The aim of periodontal regeneration therapy is to re-establish the supporting structures of the teeth that may have been damaged because of certain diseases or injury (Hägi et al., 2014). In addition, these stem cells are effective and used in the regeneration of cementum tissue and periodontal ligament (Zakrzewski et al., 2019). Cementum is classified as mineralized tissue that is similar to bone, as well as covering and protecting the root (Lin & Yelick, 2008). Ultimately, leading to the clinical importance of utilizing dental stem cells in regeneration therapy. Stem cells from the apical papilla (SCAP) are located within the apical papilla which is tissue surrounding the immature teeth (Zakrzewski et al., 2019). These specific cells are mesenchymal cells which allows for a differentiation of cell types including adipocytes (fat cells), cells like neurons (brain cells) and similar to odontoblast cells (cells that originate from the neural crest) (Zakrzewski et al., 2019). These stem cells have a unique quality of increased capability of cell proliferation, which is the process of cell growth, division and increases in cell number (Zakrzewski et al., 2019). In terms of tissue regeneration stem cells from the apical papilla would be better suited due to the cells enhanced capabilities of proliferation in comparison to dental pulp stem cells (Zakrzewski et al., 2019). It can be established that through the discovery of stem cells in the oral cavity, profound implication of cell specialization and treatment options have become viable for society.

It is important to note the accessibility of dental care differentiates between the developed and developing countries. In specific, in first world countries it has become normalized to visit dental care professionals for medical and non medical purposes. In some societies this can even be deemed as luxurious. In comparison, developing countries' individuals face barriers to accessibility for health and dental care. Oral diseases in particular impact societies in developing countries due to the barriers to accessing professional medical care (Pack, 1998). These barriers can be based on the following aspects including poverty, difficult living facilities, decreased funding from the government, limited knowledge on health education, and policies for administering dental or oral practitioners in the workplace (Pack, 1998). The allocation of resources and equipment for dental care varies between the developed and developing countries. There are even charities that focus on filling the gap of inaccessibility of health care to accommodate societal needs.

The ReSurge International, is a non for profit organization that focuses on delivering surgical care to individuals in low to middle income countries (Surgical Mission Trips | ReSurge International, n.d.) . This charity provides reconstructive care to societies through local and international mission trips with the aim to improve and increase access to patient care (Surgical Mission Trips | ReSurge International, n.d.). Their work is demonstrated on individuals who have burn victims, genetic abnormalities, congenital conditions and cancer through performing multiple reconstructive surgeries (Surgical Mission Trips | ReSurge International, n.d.). Reconstructive surgeries have the possible use of needing stem cells in the treatment process. This bridges the gap in accessing stem cells in developing countries through the facilitation of non for profit organizations. These are fine examples of non governmental organizations that are advocating and acting on the need of medicare in developing countries.

Stem cell therapy can also be applied to neurodegenerative diseases (Zakrzewski et al., 2019). One of the most important discoveries associated with neuroscience is invalidating the basis that the central nervous system of an adult is unable to undergo the process of neurogenesis (Zakrzewski et al., 2019), within adults, neurogenesis is the process and formation of neurons, known as brain cells (Kumar et al., 2019). This ultimately exemplifies the ability for cell generation at the adult level in reference to neuronal cells.

There have been numerous studies done on using cell therapy as a form of treatment for degenerative disorders (Kolios & Moodley, 2013). Several clinical and preclinical trials have demonstrated favourable results on the following conditions: heart failure, leukemia, diabetes, fibrosis, cirrhosis, and nervous system disorders (Kolios & Moodley, 2013). In addition, stem cells have been found beneficial and utilized in conditions and diseases associated with symptoms of major inflammation (Kolios & Moodley, 2013). Stem cells demonstrate the ability to aid in the treatment of several disorders, aiding members of society with treatment and care options.

Parkinson's disease is a type of neurodegenerative disease and will be discussed in its relation to stem cell therapy. Approximately 1 % of the population in regards to 65 years or greater is impacted by Parkinson's disease within North America (Master et al., 2007). Due to two decades of medical experience and clinical research on neural transplantation of human fetal tissues, there is an increased likelihood of the assessment of stem cell transplantation on Parkinson's disease (Master et al., 2007).

There are a variety of stem cell transplantation utilised in the treatment of Parkinson's disease including: fetal ventral mesencephalic cells, embryonic stem cells, induced pluripotent stem cells and neural progenitor cells (Jang et al., 2020). In particular, the focus will be understanding the effectiveness of fetal ventral mesencephalic transplants on patients diagnosed with Parkinson's. Fetal ventral mesencephalic transplants are the implantation of human fetal tissue into individuals with Parkinson's disease (Master et al., 2007).

Globally, over 350 patients have undergone fetal ventral mesencephalic tissue transplantation in a variety of study conditions including: placebo-controlled, double blind, open label, randomization of clinical trials eliciting a variety of results (Master et al., 2007). In reference to the clinical trial that implemented a placebo control on subjects, researchers observed that in 10 participants who were less than 60 years old made some improvement (Master et al., 2007). One of the other trials was unable to deem any progress in terms of motor features within the sample size of 34 subjects (Master et al., 2007). Within both trials, subjects diagnosed with Parkinson's disease whose cases were less severe demonstrated clinical improvements (Master et al., 2007).

There is one drawback that prevents tissue transplantation being an accessible and routine treatment for neurodegenerative diseases, which is the limited supply of the right kind of tissues (Master et al., 2007). Stem cells can be a form of biological material that can be used in cell therapy, when cultured under standardized conditions. Which can possibly lead to unlimited supply of stem cell based tissue to be used in transplantation therapy in reference to treating Parkinson's disease (Master et al., 2007). The utilization of stem cells in this area will facilitate restorative process and transplantation of the stem cells (Master et al., 2007). More research is required to meet the demand for stem cells that this treatment will create.

Parkinson's disease impacts the aging population. The benefits of finding a cure to Parkinson's disease will allow for an increase in longevity within the elderly populations. With the impact of stem cell based therapy and the possible implications on improving brain function like neurogenesis this could imply increasing basic physical and mental functionalities. The question is whether medical advancements in stem cells can facilitate individuals with Parkinson's disease to baseline level performance? If stem cell technology was able to find a cure for Parkinson's disease this could alleviate the burden on societal members who take care of their elderly families. This means that the elderly population has the capability to be able to work and operate as independent entities without the support of the family. Families would be able to coexist and decrease the amount of care required to take care of the elderly. Thus, it is important to analyze the implications on how stem cells could possibly correlate to the curing of neurodegenerative diseases like Parkinson's.

Through this chapter the emphasis was on understanding stem cells' impact on society. The increase in the number of stem cell transplants over the years is a prime example of the number of individuals that have benefited from the discovery of stem cells. As experimental treatments progress into clinical trials, the possibility for new forms of treatments develops and normalizes the use of stem cells in society within medicine. There is a significant amount of research to be done in order for stem cells to be accessed and standardized as a form of treatment. Stem cells can impact every avenue of society in terms of socio-economics, policies, education and medicine. It is equally important to question the purpose and the importance of studying stem cells at the micro level, specifically individuals. The significance of the study of stem cells in terms of medicine and drug testing will be analyzed and critiqued in the next chapter.

4

Why is Studying Stem Cells Important?

*Written By **Pareesa Ali***

INTRODUCTION

The previous chapters have discussed the background behind stem cells, how they were discovered, and the impact of their discovery on society. This chapter will now discuss the significance of studying stem cells and why they are an important scientific advancement. Stem cells offer the capability to treat inflammation, cell death, fibrosis and the repairment of tissues. These processes are responsible for many diseases and conditions that cause harm to an individual, and do not currently have any available treatments. In addition, stem cells have allowed for breakthroughs in the testing of pharmaceutical drugs by providing highly individualized tissues to use as models in the laboratory. Thus, the discovery and subsequent use of stem cells in medicine has had an instrumental impact in the field of medicine as a whole. First, this chapter will discuss the significance of discovering stem cells with regards to its contributions in the field of medicine for the individuals who have received stem cell therapy.

Next, we will cover the overall impact of this discovery and how stem cell therapies have changed the scope of regenerative medicine, by providing many individuals with a second chance at life. Following this, we will discuss the importance of this discovery in terms of testing new drugs and treatments for previously incurable diseases. Finally, we will outline some of the various diseases and conditions that stem cell therapies have the potential to treat, and how they have improved the lives of patients who were suffering from harsh diseases and conditions, along with a poor quality of life.

THE IMPACT OF STEM CELLS IN MEDICINE

As previously discussed in earlier chapters, stem cells from one particular organ system have demonstrated an ability to develop into the differentiated cells from another organ system, such as the liver, brain or the kidney (Alison et al., 2002). This ability of stem cells to differentiate into almost all cells has been an incredible scientific discovery (Alison et al., 2002). It has revolutionized the future of how cardiovascular diseases, neurodegenerative diseases, cancer, diabetes, and other diseases can be treated (Alison et al., 2002). Prior to the development of stem cell treatments, mortality rates resulting from malfunctioning vital organs were high in all societies (Alison et al., 2002). Moreover, aside from the potential of stem cells in restoring diseased tissues back to a healthy state, they offer additional benefits for further understanding human development (Zakrzewski et al., 2019). Several serious medical conditions, such as cancer or birth defects, are caused by a process of improper differentiation or cell division (Zakrzewski et al., 2019). Using stem cells in research can help better understand stem cell physiology, which is the first step for finding a new way to treat previously incurable diseases (Zakrzewski et al., 2019).

Moreover, in the scientific field, aging is regarded by some as a reversible epigenetic process (Zakrzewski et al., 2019). The term epigenetics refers to the study of how an individual's environment and behaviours can change and affect the way their genes work. Thus, the ability to use stem cells to implement cell reprogramming as a way to delete the negative environmental and behavioural effects that are responsible for aging have been investigated (Zakrzewski et al., 2019). Specifically, stem cell therapy is currently being investigated for delaying the progression of previously incurable neurodegenerative diseases, such as Parkinson's, Alzheimer's and Huntington's disease, which are associated with age (Zakrzewski et al., 2019). These neurodegenerative diseases are caused by the death of certain parts of the brain (Zakrzewski et al., 2019). Previously, there was no treatment that would make the brain capable of neurogenesis, which is the process of developing new cells (Zakrzewski et al., 2019). However, the discovery of neural stem cells is proving to be an incredibly significant treatment for patients of neurodegenerative diseases, as these individuals suffer from an extremely low quality of life. Preclinical studies testing stem cell therapy in rodent models have demonstrated that neural stem cells are capable of improving cognitive function (Zakrzewski et al., 2019). Additionally, beyond delaying progression of these diseases, stem cell therapy may also have the potential to remove the source responsible for causing these diseases (Zakrzewski et al., 2019). Stem cell therapy is an ideal and promising treatment as it has demonstrated that stem cells can be an important and achievable therapy to prevent some of the harmful consequences of aging. Therefore, the impact of stem cells in medicine have provided the potential for treating many diseases which are currently classified as incurable and untreatable. Overall, stem cell therapies present an opportunity for improving the lives of individuals suffering from these diseases and states, by offering them an effective and accessible treatment.

STEM CELLS & REGENERATIVE MEDICINE

Stem cells have also demonstrated a lot of potential in the field of regenerative medicine. Stem cells have provided the medical field with the opportunity to replace tissues that are worn out, either by old age or disease (Alison et al., 2002). The opportunity to use the body's own growth factors to engineer tissues and complex internal organs outside of the body is an incredible scientific advancement. In the field of regenerative medicine, stem cells have been widely explored in preclinical and clinical trials. These trials are focused on the reconstruction of fragile tissues, including tissues found in the musculoskeletal system, nervous system, myocardium, which is the muscular tissue of the heart, and the liver, cornea, trachea, and even the skin (Han et al., 2019). In particular, mesenchymal stem cells can differentiate into almost any type of cell, and they are an ideal source for cell tissue regeneration (Han et al., 2019). This is due to the fact that the properties of mesenchymal stem cells revolve around anti-inflammatory, immunoregulatory, and immunosuppressive capacities (Han et al., 2019). The extraction of these cells, followed by their differentiation and immunological properties, make them an ideal therapeutic in regenerative medicine (Han et al., 2019). Additionally, these stem cells modulate immune responses, without a substantial risk of immune rejection (Patel et al., 2013).

Stem cells have been used in medicine as regenerative therapies for disease and they have even demonstrated potential for the treatment of injuries. A common injury seen in professional sportsmen is an injury to the tendon, known as tendinopathy (Zakrzewski et al., 2019). Current treatment for tendon injuries focuses on conservative or surgical treatments, which aim to manage and heal the tendon (Zakrzewski et al., 2019). However, these often do not lead to optimal results and outcomes, as surgery can have a high risk of rerupture, even after treatment (Zakrzewski et al., 2019). Moreover, the reason behind tendon injuries relates to their lack of regenerative capabilities (Zakrzewski et al., 2019). After an injury, tendons heal by forming scar tissue, as they are incapable of regenerating after an injury (Zakrzewski et al., 2019). This scar tissue lacks the functionality seen in healthy tissues, due to reasons such as pain or swelling (Zakrzewski et al., 2019). However, the regenerative potential of stem cell therapy can have a significantly positive effect on tendon healing (Ahmad et al., 2012). The regenerative effects of stem cells can stimulate and produce tendon tissue that is similar to its pre-injurious state (Ahmad et al., 2012). This potential for tissue engineering through stem cell therapy is major, not only in the medical field, but also in the field of sports, as many injured players can have the potential to return to their original healthy states after obtaining a sports injury. Without stem cell therapies, these individuals suffer from major complications after surgery, or from a lower standard of living by having to resign from their profession as a result of the injury. Stem cell therapy offers these individuals a successful chance at injury repair, without the potential for complications and recurrence.

IMPORTANCE OF STEM CELLS IN DRUG TESTING

Studying stem cells in the laboratory has also allowed scientists to learn about essential properties of stem cells and how they differ from other specialized cell types (Importance of Stem Cells). Studying these cells in the lab presents an opportunity to screen new drugs under development, and to develop model systems to study normal growth in humans (Importance of Stem Cells). Stem cells have allowed for the ability to create personalized treatments for patients, and to understand diseases and how a treatment will affect a patient. Thus, experiments for new drugs requiring testing on living tissue can be performed safely and effectively on specific stem cells (Zakrzewski et al., 2019). This reduces the potential for harmful side effects that may be observed on human study participants. If any undesirable effect appears, the formula for the drug can continually be changed until the drug is safe for use (Zakrzewski et al., 2019). Currently, more than 90% of drugs that are tested in clinical trials fail to be approved for market use (Rubin, 2008). This places a major burden on the health care system, as time and money is invested into testing these compounds which fail the efficacy or toxicity standards (Rubin, 2008). However, using stem cells for drug testing may rectify these inefficient and costly issues, by providing a model that better represents the diseases seen in humans (Rubin, 2008).

Stem cell models can be used to study and understand the underlying mechanisms behind a disease, in order to develop effective and validated treatments. By testing treatments for diseases on the stem cells of patients with a specific disease, drug discovery can become more efficient at discovering and testing therapeutics. (Rubin, 2008). Additionally, in the early stages of drug discovery, high-throughput screening technologies are used prior to clinical trial stages (Ebert

& Svendsen, 2010). These screening technologies rely on compound libraries made up of cell lines, which are used to test a variety of drugs and compounds under development (Ebert & Svendsen, 2010). However, these cell lines do not provide an accurate indication of how these drugs affect human cells, leading to the potential for low efficacy (Ebert & Svendsen, 2010). Clinical trials using volunteers also put these patients at risk when testing a new drug. Therefore, using stem cells that can be differentiated into any cell type can provide a better alternative for testing the efficacy and toxicity of a drug, providing similar results as expected in humans (Ebert & Svendsen, 2010). Overall, this can have major implications for individuals, as it reduces the need for unnecessary human trials which are testing drugs that have the potential for risky side effects. In addition, investigating and testing treatments on stem cells also improves the likelihood of developing a drug with a higher chance of success for the patient. Furthermore, drug testing on stem cells can impact and improve the healthcare field by increasing confidence in the safety and efficacy of drugs and treatments that are administered to the patient, which is a major benefit for those who are undergoing the treatment (Matsa et al., 2014).

IMPACTS ON THE INDIVIDUAL

Stem cell therapy in animal and human clinical trials has shown promising results, especially for treating those individuals who suffer from a disease without an effective or easily available treatment or cure. These diseases and conditions include hematologic diseases, type 1 diabetes, cardiac disease, and patients requiring organ transplants, among many other disorders. To begin, stem cell therapies have offered individuals suffering from hematologic diseases a second chance at life. Hematologic diseases include acute and chronic conditions that are caused by blood disorders, such as multiple myeloma and leukemia (Xu et al., 2013). Several of these conditions can be life-threatening, and do not currently have a cure. Conventional treatments for these diseases include chemotherapy and radiation, which come with a myriad of harmful side effects (Xu et al., 2013). However, the use of stem cell transplantation can provide a more effective and less toxic alternative to treatment for these diseases (Xu et al., 2013). This can provide patients with a treatment that has a higher chance of success, as well as an overall improvement to their quality of life.

Moreover, the development of stem cell treatments has also impacted individuals who are waiting for organs by reducing the need for patient organ transplants (Hoogdujin et al., 2010). Mesenchymal stem cells are showing great efficacy in organ transplantation, without the need for patient donors. This is an incredible opportunity for individuals who are on organ transplant wait lists for months, and even years. Stem cell therapy in place of organ transplants can offer multiple benefits to various individuals; from increasing the life-span of deteriorating organs, to treating patients whose bodies have rejected organ transplants in the past (Hoogdujin et al., 2010). Another condition benefiting from this discovery is type 1 diabetes. Diabetes is a disease that results from the destruction of insulin-producing cells in the pancreas due to an autoimmune reaction (Zakrzewski et al., 2019). Stem cell therapy offers the potential for treating this disease by inducing stem cells within the body to differentiate into insulin-producing cells, in order to offset diabetes (Zakrzewski et al., 2019).

In addition, heart failure is a common condition resulting from various cardiac disorders, especially in adults (Xu et al., 2013). Cardiomyocytes, which are the heart muscle cells, have a very limited ability to regenerate (Xu et al., 2013). This makes it extremely difficult for the heart to recover after injury or a cardiac episode, thus greatly limiting the potential for treating cardiac diseases (Xu et al., 2013). However, stem cell therapy for patients with heart failure is offering a successful future breakthrough in this field. Stem cells can be used to reprogram cardiomyocytes so they are capable of healing and repairing damage caused by heart diseases in the future. Overall, these conditions are all highly prevalent in the community, yet there is no effective treatment for curing patients of these disorders. The prospect of stem cell therapy can reduce this burden of disease by providing these patients with a shot at recovery and an enhanced quality of life.

CONCLUSION

This chapter has discussed the significance of stem cells in science, with a focus on how this discovery has impacted the lives of individuals suffering from a variety of conditions. The impact of stem cells has been significant in the field of medicine, from the use of stem cells in research to better understand individual diseases, to using stem cells to further develop and understand potential treatments for these diseases. This chapter also discussed the significance of stem cells in transforming the field of regenerative medicine, and how this discovery can impact the lives of professional athletes who suffer from injuries that would have previously been classified as career-ending. Next, this chapter discussed the importance of stem cells in drug testing as a way to work towards treatments that have a high efficacy and low toxicity for patients. Finally, we examined the various ways in which stem cells have impacted individuals by improving the quality of life for many individuals suffering from chronic and malignant conditions. All of these capabilities offered by stem cells, from regenerating damaged tissues, to testing novel therapeutics, can have major impacts for individuals suffering from diseases and a standard of living. The upcoming chapters will discuss stem cells and the specific stem cell treatments currently being investigated in more detail. Additionally, the chapters will also cover the science behind this novel therapeutic and the future directions arising from this major discovery.

5 What are Stem Cells?

*Written By **Si Cong (Sam) Zhang***

INTRODUCTION

In recent years, "stem cells" have become a buzzword, thrown around on television, social media, and newspapers. It seems like every other day one hears about the great powers of stem cells, how they will cure a certain disease or reverse aging. As information gets passed from the scientist to scientific journals, scooped up by a science reporter, reviewed by an editor, published, and reposted by the general public on social media, the context is often lost. In the end, to most people "stem cells" become the real-life equivalent of the "MacGuffin", a certain object that advances the plot, but irrelevant in itself. As stem cell applications inch closer to reality, it becomes more necessary for people to understand stem cells to avoid irrational fears and myths surrounding this sensitive subject.

ETYMOLOGY

First of all, what does the word "stem cell" actually mean, and where does it come from? It might be surprising to learn that the word "stem cell" has had a variety of meanings formerly associated with it.

A stem cell is a kind of cell, so that prompts the question: what is a cell? The first application of the word "cell" to describe a microscopic biological unit is made by the English scientist Robert Hooke in his book "Micrographia". Micrographia contains a wide variety of biological observations under a microscope, including a thinly sliced piece of cork. In Micrographia, Hooke states:

...I could exceeding plainly perceive it to be all perforated and porous, much like a Honey-comb, but that the pores of it were not regular; yet it was not unlike a Honey-comb in these particulars.

First, in that it had a very little solid substance, in comparison of the empty cavity that was contain'd between, as does more manifestly appear by Figure A and B of the X I. Scheme, for the Interstitia, or walls (as I may so call them) or partitions of those pores were neer as thin in proportion to their pores, as those thin films of Wax in a Honey-comb(which enclose and constitute the sexangular cells) are to theirs.

Next, in that these pores, or cells, were not very deep, but consisted of a great many little Boxes, separated out of one continued long pore, by certain Diaphragms, as is visible by the Figure B, which represents a sight of those pores split the long-ways.

I no sooner discern'd these (which were indeed the first micro-scopical pores I ever saw, and perhaps, that were ever seen, for I had not met with any Writer or Person, that had made any mention of them before this) but me thought I had with the discovery of them, presently hinted to me the true and intelli-gible reason of all the Phænomena of Cork; As,
(Hooke, 1665, p.113)

Soon after, Leeuwenhoek would become the first person to see living cells using a more powerful microscope, finding bacteria and protozoa which he called "animalcules", and other cells such as blood cells and sperm (Gest, 2004). Since then, people have looked at objects and organisms under more and more powerful microscopes, learning more and more about cells in the process. But the modern cell theory would only begin to take form in 1838 when Theodor Schwann proposed its first two tenets:

- All living organisms are composed of one or more cells
- The cell is the most basic unit of life

But he was wrong on one aspect which is critical to understanding stem cells. He proposed free cell formation, where cells are created when materials assemble into a cell spontaneously. This was proven wrong by Rudolf Virchow who correctly recognized the third tenet:

- All cells arise only from pre-existing cells (Schultz, 2008)

While the modern version has included other parts such as activity, metabolism and heritability, these three tenets remain largely unchanged, and it is impossible to know what a stem cell is without a good understanding of these three tenets.

While most people know what a cell is to some degree, few know why a stem cell is called a "stem" cell. While the stem cell does not have any physical relation to the stem of a plant, it is related in an abstract sense. The first instance of the word "stem cell" being used is in fact addressing something entirely different from the modern meaning. The German biologist Ernst Haeckel uses the German word for family tree "Stammbäume" (literally "stem tree") in the context of the tree diagrams of species with common ancestors in his book "Natürliche Schöpfungsgeschichte: Gemeinverständliche wissenschaftliche Vorträge über die Entwickelungslehre im Allgemeinen und diejenige von Darwin, Göthe und Lamarck im Besonderen, über die Anwendung derselben auf den Ursprung des Menschen und andern damit zusammenhängende Gründfragen der Natur-Wissenschaft" or in English "Natural history of creation : Commonly understood scientific lectures on the theory of evolution in general and that of Darwin, Göthe and Lamarck in particular, on the application of the same to the origin of man and other related fundamental questions of natural science." regarding evolution.

Specifically, the context in which the word "Stammzelle" (literally "stem + cell") is used is to address the ancestral single-celled organism from which all life evolved, now called the LUCA or Last Universal Common Ancestor (Haeckel, 1868).

It is not unreasonable for one to wonder why stem cell no longer means LUCA even though the word made perfect sense in that specific context, at least in German. This is due to the fact that Haeckel proposed that "Stammzelle" also means a fertilized egg in the context of embryology. Just as LUCA gave rise to all other organisms, the fertilized egg gives rise to all other cells in an organism (E. H. P. A. Haeckel, 1877). The use of the word "Stammzelle" was expanded by Theodor Boveri as being all the cells which lie directly in between the developmental path of a fertilized egg and a sex cell(sperm or unfertilized egg) (Boveri, 1892). Valentin Häcker, a colleague of Boveri also used the word to describe a large cell which divides into a cell which becomes the mesoderm and another which becomes the germ cells (Häcker, 1892). All three of these definitions would fall under what are now called "germline cells".

The first use of the English word "stem cell" is by William Sedgwick to describe the parts of a plant which grows and regrows in 1886. (Richardson, 1998) However, the word was made famous by the book "The Cell in Development and Inheritance" written in 1896 by Edmund B. Wilson. This book discussed the works of Boveri and Häcker, therefore stem cells referred to germline cells (cells that eventually becomes a sex cell) in the book.

Simultaneously in 1896, when searching for the progenitor of all blood cells, Artur Pappenheim also used "Stammzelle" to describe such a cell. Today, such a cell is known as a hematopoietic stem cell.

The modern definition of stem cells retained the self-renewal and differentiation aspects of the original definitions and expanded to contain any kind of cell which can do both. This includes a very large category of cells with varying ability to differentiate and self-renew. In general, stem cells can be divided into 3 groups: embryonic stem cells, adult or somatic stem cells, and Induced Pluripotent Stem (IPS) cells.

In general, as the cells develop from the original single cell zygote, they become more and more specialized in a specific role and the range of cells they can potentially become decreases. The order goes from totipotent stem cells which can become any human cell including the placenta and amniotic sac, to pluripotent which can become any cell of the human body, to multipotent which can only become one of a few related cell types. This process of specialization begins almost immediately after fertilization.

EMBRYONIC STEM CELLS

When the egg is fertilized, it begins to divide in half, again and again for a total of 4 times, into 16 cells in a spherical clump. This clump is called the morula, and every cell in the morula is a totipotent stem cell. Meaning that none of these cells are specialized and can become any human cell. (Forgacs & Newman, 2005)

Four to five days after fertilization of the human egg, the embryo undergoes multiple divisions and results in 150-200 cells forming a hollow sphere. This process of forming a hollow sphere is called blastulation, and the resulting hollow sphere is called the blastocyst. (Forgacs & Newman, 2005) The blastocyst consists of the outer layer of cells (trophoblast), the inner cavity (blastocoele), and the inner cell mass (embryoblast). At this point, the cells of the blastocysts have already been specialized in order to develop into different parts of the embryo. The trophoblast will eventually develop into the placenta, and the inner cell mass will become the fetus. (Forgacs & Newman, 2005)

The embryonic stem cells in question are cells derived from this inner cell mass which will become the fetus. Embryonic stem cells are therefore pluripotent, meaning that they have the capability to develop into any cells of the human body. (Thomson et al., 1998) Because of the pluripotency of embryonic stem cells, there is significant research being done on the therapeutic potential of embryonic stem cells. However, there are also debates within the scientific community regarding the ethics of using embryonic stem cells as removing the inner cell mass results in the death of the embryo. Both of these topics are discussed in chapter 6 and chapter 9 respectively.

ADULT STEM CELLS

Further into embryonic development, when the cells of the inner cell mass have further divided and specialized, though the detailed mechanism is unclear, eventually some of the cells of the blastocyst become adult stem cells. Adult stem cells, also called somatic stem cells, are multipotent stem cells which can be found across various systems and throughout the entire body. The task of adult stem cells is to replace and replenish dying cells via cell division and differentiation (4. The Adult Stem Cell | Stemcells. Nih. Gov, n.d.).

While they can be found in most organ systems, adult stem cells are located in very specific regions of the organs called niches. One example of such a niche is the basal layer of the skin.Under normal conditions, the adult stems remain mostly inactive. But when there is a change in the niche, such as no longer being in physical contact with adjacent cells due to an injury, the stem cell undergoes cell division and cell differentiation to replenish the lost cells and restore the niche. The cell division will stop once the niche is restored and the stem cell will again enter into quiescence (an inactive state in which the stem cell waits for the next time the niche is disturbed) (So & Cheung, 2018). This process of controlled cell division is extremely important for maintenance of the human body, as many kinds of cancers are caused by the uncontrolled division of adult stem cells (O'Connor & Adams, 2010).

Unlike an embryonic stem cell which can produce any cell type in the body (pluripotent), the adult stem is more constrained in its ability to produce different kinds of cells. Usually, the adult stem cell can only produce several usually related types of cells (multipotent) (Mummery et al., 2014). For example, the hematopoietic stem cell, a kind of stem cell which resides in the bone marrow, is responsible for making various kinds of blood cells, but it is unable to produce other kinds of cell. While the current general consensus that the adult stem cell is limited in its ability to divide into different kinds of cells, this idea has been recently challenged by findings that suggest stem cells can sometimes produce other cell types beyond those of the tissue from which the stem cells were taken from. This new idea is called adult stem cell plasticity, and is still being investigated (Chatterjee et al., 2010).

When the adult stem cell divides, it can either divide symmetrically, or asymmetrically. Symmetric cell division results in two stem cells identical to the original cell, whereas asymmetric cell division results in one stem cell and one progenitor cell (Shahriyari & Komarova, 2013). The progenitor cell or precursor cell is a partly differentiated cell which is only able to produce certain types of fully differentiated cells (Clayton et al., 2007). This process is easier to explain using the previous example of hematopoietic stem cells. The hematopoietic stem cell is the adult stem cell, and it is able to undergo asymmetric division to produce a progenitor cell. There are two kinds of progenitor cells produced by the hematopoietic stem cell, the lymphoid progenitor cell and the myeloid progenitor cell. The lymphoid progenitor cell is committed to producing immunity related blood cells such as B cells, T cells, and natural killer cells. Myeloid progenitor is committed to producing neutrophils, macrophages, platelet producing cells, and red blood cells. Progenitor cells also have the ability to undergo symmetric division to produce more of themselves (5. Hematopoietic Stem Cells | Stemcells.Nih.Gov, n.d.).

As stated earlier, in general, a pattern can be seen where cells become increasingly specialized and committed to becoming a specific cell type over time. In nature, this process is usually irreversible, meaning a specialized cell cannot go back to a previous stage. Of course, it's not exactly that simple and recent research has shown that some cells can go back to specialize to something else. However, discussions of this research go beyond the scope of this chapter.

INDUCED PLURIPOTENT STEM CELLS

The same, however, cannot be said when human intervention is added to this natural process. Induced pluripotent stem cells are essentially adult stem cells that have been reprogrammed by scientists to be pluripotent. As indicated in chapter 2, induced pluripotent stem cells are a relatively new invention, only made possible by recent advances in DNA technology.

The principle behind induced pluripotent stem cells is that the differentiation and specialization seen in all cells are caused by epigenetic expression of certain genes and the inhibition of others. There are four specific genes (Oct3/4, Sox2, Klf4, c-Myc) encoding for proteins which are normally present in high quantities in embryonic stem cells but not adult cells. Through experimentation, scientists found it is possible to "revert" the adult stem cell to a state similar to that of the embryonic stem cell (Takahashi & Yamanaka, 2006). These four transcription factors are collectively named Yamanaka factors after Shinya Yamanaka who pioneered this technology.

Because induced pluripotent stem cells can be sourced from the patient, they are genetically identical to the patient, having much greater research and therapeutic potential. This point is talked about in more detail in chapter 6. While they possess significant therapeutic potential, there is a very controversial and potentially dangerous aspect of induced pluripotent stem cells. That is the gene c-Myc used to produce induced pluripotent stem cells, is also a proto-oncogene. Meaning a higher than normal expression of the c-Myc gene could potentially result in cancer (Okita & Yamanaka, 2011). This is one of many major barriers for induced pluripotent stem cells to overcome if it will ever become mainstream.

CORD BLOOD STEM CELLS

While the clinical application of many stem cell treatments are still in its infancy, one particular practice is rapidly gaining popularity. Cord blood is the blood which remains in the placenta and the umbilical cord after childbirth. This blood contains multipotent stem cells which can be stored at a cord blood bank for use in the future. Cord blood stem cells present a cheap source of somatic stem cells which have the same genetic information as the child, thereby presenting no potential transplant rejection (Weiss & Troyer, 2006). More and more parents are preserving cord blood for their children, and this could affect the availability of stem cells in the future (Shearer et al., 2017).

CONCLUSION

In summary, while stem cells may appear to be a challenging topic to tackle, it is easier and more approachable in reality. To anyone with interests in biology or wishes to get a better understanding of the situation, it can be a very interesting and relevant topic to learn more about. Not only will it improve understanding of certain medical advancements, it could also provide insight into sensitive topics such as whether a blastocyst should be regarded as a human life.

6

What are Prominent Stem Cell Therapies & Treatments Used Today?

*Written By **Salma Abrahim***

INTRODUCTION

With advancing research, stem cell therapy has become a prominent area of interest, particularly in the area of regenerative medicine. The ability to generate different types of tissues and cells from stem cells has evoked great debate and interest. This chapter will delve into prominent stem cell treatments that are currently available as well as some that are in the research phase and have the potential to be implemented in the future. First, general information regarding pluripotent and multipotent stem cells will be discussed, and then the chapter will delve into specific treatments including bone marrow transplants, alternatives to arthroplasties, epigenetic reprogramming, therapy for Type 1 diabetes mellitus, stem cells from human exfoliated deciduous teeth (SHED) and finally, comparing pluripotent and multipotent cells as therapeutic agents.

PLURIPOTENT STEM CELLS

The therapeutic usage of pluripotent cells is not as prominent as multipotent cell usage. As early animal-based studies have shown, pluripotent stem cell therapy may cause the formation of tumors called teratomas, and therefore they have not been used in therapeutic treatments for humans (Biehl & Russell, 2009). The teratomas are made out of a mix of cells from the early germ layers, which are the first cell layers found in the early stages of embryonic development (Biehl & Russell, 2009). But, pluripotent stem cells have been used to treat animals with various conditions including diabetes, by creating cells that are insulin-producing and are responsive to increased glucose levels in the blood (Biehl & Russell, 2009). In addition, animals with visual impairments or various spinal cord injuries have been previously treated with new retinal epithelial cells for their vision and new myelinated neurons for their spinal injuries (Biehl & Russell, 2009). Ongoing animal studies are being conducted to explore the effectiveness of pluripotent stem cells to treat Parkinson's Disease, heart failure and muscular dystrophy (Biehl & Russell, 2009). For example, there is a possibility that stem cells can be transplanted and produce dopamine for curing Parkinson's disease (Wernig et al., 2008). Particularly for cardiac function, stem cell therapy has the ability to integrate newly formed cardiac myocytes (muscle cells that generate electrical impulses in the heart) into the myocardium to increase beating force (Biehl & Russell, 2009). Studies have shown that cardiac myocytes that come from human embryonic stem cells are able to form viable human myocardium muscle after stem cell transplantation into animals, with some evidence of electrical integration (Biehl & Russell, 2009). Damaged rodent hearts showed cardiac function improvement after the injection of human-derived cardiac myocytes (Biehl & Russell, 2009). The mechanism of the resulting increased cardiac function is not fully known, and further research is needed (Biehl & Russell, 2009).

A second type of pluripotent stem cell therapy that is currently being researched is somatic cell nuclear transfer (ntES cells) (Pluripotent, 2021). This process involves transferring the nucleus, which contains the genetic information, from a somatic or body cell to an unfertilized egg cell (Pluripotent, 2021). The transfer of a different nucleus into the donor egg "reprograms" the cell into a pluripotent cell, reactivating the genes needed to make all tissues in the body (Pluripotent, 2021).

The egg develops and produces pluripotent stem cells, which can be extracted from it (Pluripotent, 2021). This process has been successful in animals, and to make these cells in humans, an egg donor and a skin cell from the patient is needed (Pluripotent, 2021). If proved successful in humans, ntES cells would eliminate tissue rejection and matching issues that are prevalent with stem cell therapies (Pluripotent, 2021)

Similar to ntES cells, induced pluripotent stem cells (iPS) are "reprogrammed" cells that have been converted to pluripotent cells from an ordinary cell through a gene transfer (Pluripotent, 2021). These cells are a promising approach for cell therapy in many diseases that have no cures, such as heart infarction and diseases of the kidney, as they are able to provide "new" cells to mediate the disease (Moradi et al., 2019). As iPS cells are produced from the patient's own cells, they can be used to avoid rejection complications by the immune system of the patient (Moradi et al., 2019). It is important to note that in several animal studies, there were some viruses used to introduce genetic reprogramming that were seen to cause cancer, and therefore further research is needed to understand how iPS can be used as a therapeutic tool (Pluripotent, 2021).

Lastly, with chemical treatments, eggs that are unfertilized can be "tricked" to become embryos without the fertilization process initiated by a sperm, a procedure called parthenogenesis (Pluripotent, 2021). After the embryos develop and mature in a lab, pluripotent stem cells can be taken from them (Pluripotent, 2021. There is a future possibility that a woman can donate her own eggs to make parthenogenetic embryonic stem cells (pES) that match her genetic makeup which her immune system will have a lower chance of rejecting the stem cells (Pluripotent, 2021). In addition, through the process of genetic typing and engineering, it is possible to create treatments using pES for other patients other than the donor herself, through creating stem cells banks to match different tissues (Pluripotent, 2021). This process has been proven to work in animal-based research, however more research is needed to investigate if tissues derived from this process can function normally (Pluripotent, 2021).

MULTIPOTENT STEM CELLS

Since the 1960's, multipotent stem cells have been harvested from bone marrow to treat different diseases and conditions, including myeloma and leukemia (Biehl & Russell, 2009). These cells are able to give rise to erythrocytes and lymphocytes, and therefore are able to treat various blood cancers (Biehl & Russell, 2009). Bone marrow transplantation has also been used to treat other conditions, for example forming whole joints in animal models in research (Biehl & Russell, 2009). This has been achieved using mesenchymal stem cells which can differentiate into cartilage and bone cells (Biehl & Russell, 2009). Currently, there are clinical trials being conducted to test if stem cells from bone marrow can treat heart ailments (Biehl & Russell, 2009). In addition, an important form of multipotent stem cells that current research is investigating is neural cells that differentiate into nerve cells (Moradi et al., 2019). By isolating these cells from fetal and adult brain tissues, and differentiating into new nerve cells that can replace damaged cells, this can be a remedy for spinal cord injuries (Moradi et al., 2019).

BONE MARROW TRANSPLANT

Bone marrow is fatty tissue located inside bones, and it creates several parts of blood including platelets, red blood cells and white blood cells (Krans, 2018). Marrow also contains hematopoietic stem cells (HSCs) which are blood-forming stem cells (Krans, 2018). This is a critical part of bone marrow as HSCs have the ability to differentiate into several different types of blood cells throughout their lifespan because they are unspecialized (Krans, 2018). A bone marrow transplant is a procedure that replaces bone marrow that has been damaged by a disease, chemotherapy or an infection with healthy bone marrow (Krans, 2018). The procedure entails the transplantation of blood stem cells that travel to the host's bone marrow, where they are able to produce and increase the rate of new marrow growth (Krans, 2018). These cells can help the host's immune system fight and avoid infections, anemia, and other blood disorders that occur with different conditions (Krans, 2018). The healthy stem cells needed can come from the host's own body where cells are grown, harvested and stored before starting a radiation treatment (as cells are not damaged yet), or from an eligible donor (Krans, 2018).

There are two different types of bone marrow transplants which can be used, depending on the reason the patient needs the procedure (Krans, 2018). The first type is autologous transplants, which use harvested stem cells from the patient, taken before they start a cell-damaging treatment (Krans, 2018). After the treatment is terminated, the patient's healthy cells are returned (Krans, 2018). Autologous transplants are not a viable option in every situation, as viability depends on the patient's bone marrow health, but they reduce the risk of graft versus host disease (GVHD), where donated bone marrow can attack the host's body as the cells view the body as foreign (Krans, 2018). The second type is allogeneic transplantation, where cells are used from a donor that has a genetic match that is close to the host (Krans, 2018). Blood relatives are often compatible donors as their genetic composition closely matches the hosts, but donor registries can have viable individuals as well (Krans, 2018). Success of autologous transplants are dependent upon the similarity between the host and donor's cells and genetic makeup (Krans, 2018). If the patient undergoes a cell-damaging treatment or has a condition that damages their bone marrow cells, allogeneic transplants are a viable option, although they have a higher risk of complications than autologous transplants (Krans, 2018). Complications can include GVHD, mucositis (inflammation of the lining of the GI tract or mouth) and graft failure, where the transplanted graft is rejected by the host's body (Lalla et al., 2008). During this type of transplant, the patient will be on immune-suppressing medications to decrease the chance of the host's body attacking the new cells (Krans, 2018).

Stem cells are collected through a few methods (Krans, 2018). For an allogenic transplant, cells are harvested from the donor a few days before the procedure is planned (Krans, 2018). Cells are harvested from the donor's hip bones with a needle (Krans, 2018). The second method through which stem cells can be harvested is through leukapheresis, where five shots are administered to the donor to help stimulate the stem cells to migrate from the bone marrow to the bloodstream (Krans, 2018). Through an intravenous (IV) line, blood is then drawn and white blood cells are separated from the rest of the blood components through a filtration machine (Krans, 2018). The donor has a central venous catheter (a tube inserted in a vein in the arm or chest) positioned on the upper right portion of their chest, allowing the fluid from the donor flow to the heart directly and spread out to the rest of the body (Krans, 2018). The new stem cells will arrive at the recipient's bone marrow and begin to grow (Krans, 2018). The transplant is completed over several days, and the catheter is left in place for the multiple sessions needed, as this provides more chances for the stem cells to integrate into the bone marrow (Krans, 2018). This process is called engraftment (Krans, 2018). Rising white blood cell count in the host's body is the first sign of engraftment, and it shows that the transplant was successful as the transplanted marrow has begun making new cells (Krans, 2018).

Recovery for a bone marrow transplant is approximately three months, but it can take a year to recover fully (Krans, 2018). The road to recovery depends on several factors including what condition is being treated, the amount and frequency of chemotherapy administered, the donor match and where the marrow transplant was performed (Krans, 2018). Some symptoms developed during the procedure can stay with the patient for the rest of their life (Krans, 2018). In a study by Pond et al., 62% of patients that have undergone a bone marrow transplant have survived a year after, and 31% of long-term survivor patient deaths were a result of reasons unrelated to the transplant (Pond et al., 2006).

STEM CELLS USED AS AN ALTERNATIVE TO ARTHROPLASTY

A big fear of many professional athletes is getting an injury, as they can be career-ending (Zakrzewski et al., 2019). Specifically, tendon injuries,which have mainly surgical treatments, don't normally have great patient outcomes (Zakrzewski et al., 2019). Tendons have a weak capability to regenerate, as they heal by making scar tissue that does not function normally as healthy tissues (Zakrzewski et al., 2019). Symptoms relating to failed healing include swelling, deposition of calcific materials and pain (Zakrzewski et al., 2019). In addition, there is a high chance of getting osteoarthritis (OA) after a tendon injury because of the lack of vasculature surrounding tendons and the lack of ability to heal (Zakrzewski et al., 2019). Arthroplasties are a common treatment for OA, but they are not an ideal procedure for younger patients as they will outlive the original arthroplasty transplant and will need additional procedures in the future (Zakrzewski et al., 2019). Stem cell therapy can help with the onset of OA, but the long-term outcomes need more extensive research (Zakrzewski et al., 2019).

CELL PROGRAMMING AND REJUVENATION

Cells from aging individuals have high levels of oxidative stress, shorter telomeres on their genes and more dysfunctional mitochondria than younger cells (Zakrzewski et al., 2019). A hypothesis was formed that if human somatic cells are reprogrammed back to iPSCs, their epigenetic (which is their gene expression) age is reset to zero (Zakrzewski et al., 2019). The epigenetic model states that at the moment of fertilization, any parenteral aging is eradicated from the zygote's genome (Zakrzewski et al., 2019). In a study by Ocampo et al., they found that the OSKM genes (which are composed of four transcription factors that reprogram somatic cells to iPSCs) can be used for regenerative purposes on affected pancreas and skeletal muscle cells (Zakrzewski et al., 2019). A big challenge in their method was using an approach that does not use transgenic animals, which are species that have had a foreign gene imputed into their genome (Zakrzewski et al., 2019). In the future, researchers want to focus on perfecting this preventative aging approach, and further delve into the possible rejuvenation of elderly individuals (Zakrzewski et al., 2019). Ethical issues of overpopulation of individuals will need to be addressed (Zakrzewski et al., 2019).

STEM CELL THERAPY FOR TYPE 1 DIABETES MELLITUS (T1DM)

The injection of exogenous insulin is not able to imitate endogenous insulin made from a healthy pancreas, but islet cell transplants have emerged as promising treatments for regulating glucose levels in patients with T1DM (Chen & Zou, 2020). In a study by Shapiro et al., seven patients with T1DM had insulin independence after the islet mass transplantation and immunosuppression, and did not need to take any exogenous insulin (Chen & Zou, 2020). In addition, glycemic control was tightly controlled in all seven patients, and with over two decades of research and improvements 60% of patients with T1DM were not dependent on insulin five years after the islet transplantation (Chen & Zou, 2020). This is important because with no dependence on insulin, islet cell transplants can be an effective method to manage T1DM. To generate insulin producing cells, researchers use iPSCs, ESCs, differentiated cells from tissues that are mature and can be transdifferentiated into IPCs and adult stem cells (Chen & Zou, 2020). After transplanting the insulin producing cells made from stem cells, they will respond to different levels of blood glucose and are able to provide enough insulin to reverse hyperglycemia over time (Chen & Zou, 2020).

There are several studies that report that bone-marrow derived stem cells (BMDSCs) also have the potential to differentiate into insulin producing cells (Chen & Zou, 2020). A study by Tang et al. concluded that BMDSCs are able to differentiate into islet-like cell clusters (ICCs) when paired with high concentrations of glucose (Chen & Zou, 2020). These ICCs can express pancreatic lineage genes including glucose kinase, GLUT2 and are able to respond to high glucose levels and release insulin when needed (Chen & Zou, 2020). The pancreatic lineage genes aid with the controlling blood glucose levels (Chen & Zou, 2020). Nevertheless, BMDSC efficacy rate is considered low and very variable (Chen & Zou,

2020). Specifically, compared to healthy adult beta cells, the amount of insulin that is secreted by BMDSCs is far less, and the expression of pancreatic lineage genes that are made by generated IPCs are much lower to the expressed genetic levels in human iselt (Chen & Zou, 2020).

STEM CELLS FROM HUMAN EXFOLIATED DECIDUOUS TEETH (SHED)

Dental pulp, a connective tissue that is highly vascularized and is encapsulated by dental and enamel, is a source of stem cells from human human exfoliated deciduous teeth (SHED) (Rosa et al., 2016). From SHED, pluripotent stem cells that are highly proliferative can be retrieved (Rosa et al., 2016). SHED can differentiate into many different cell types including osteoblasts, adipocytes and neuronal cells, and its high plasticity ability (the ability of stem cells to switch to different cells) makes it a great candidate for stem cells research in different areas (Rosa et al., 2016). Stem cells from SHED can only be retrieved from primary teeth (first teeth that are apparent in the oral cavity) that are available in postnatal human tissues (Rosa et al., 2016). In addition to its ability to differentiate into different types of cells, SHED can also undergo osteogenic differentiation (differentiating into different types of bone cells) and is able to make bone in vivo, which is a great tool for bone tissue regeneration (Rosa et al., 2016). A study by Zheng et al. highlighted this ability by having SHED mixed with a phosphate carrier, where it was found to regenerate jaw bones in swines and, when mixed with a platelet-rich plasma, it was able to form vascularized bones in dog mandible (Rosa et al., 2016).

In terms of neuronal cells, SHED mixed with epidermal growth factor (a protein that stimulates cell differentiation) had high levels of expression of neuronal markers including glutamic acid decarboxylase (which facilitates the decarboxylation of glutamate to GABA, a brain neurotransmitter) after a period of four weeks (Rosa et al., 2016). The ability of SHED to be used in regenerating different cells in vivo opens potential avenues for treating different types of conditions including spinal cord injuries, Alzehimers and focal cerebral ischemia (Rosa et al., 2016)

PLURIPOTENT VS. MULTIPOTENT

Therapeutically, the use of multipotent and pluripotent cells has different advantages and disadvantages. For pluripotent cells, they have the ability to differentiate into any cell type and they are able to treat and reverse aging tissues, which multipotent cells are not able to do (Biehl & Russell, 2009). In addition, the rate of proliferation for pluripotent cells is much higher than multipotent cells, and therefore they are able to yield a higher amount of healthy cells when compared to multipotent cells (Biehl & Russell, 2009). On the other hand, the use of multipotent cells would not require the use of immunosuppressants, as the procedure uses the host's own stem cells unlike the use of pluripotent cells, which requires immunosuppressants (Biehl & Russell, 2009). Multipotnent cells have the host's surface proteins present on the cells, and the immune system does not react to it as a foreighn body (Biehl & Russell, 2009). In terms of ethicality, multipotent stem cells do not have issues relating to ethics (as they are mainly harvested from adult patients) unlike pluripotent cells, as problems surrounding isolating fetal stem cells arise (Biehl & Russell, 2009).

CONCLUSION

With research advancements in stem cell treatments, and ongoing ethical debates surrounding these therapeutic tools, there are several options regarding how stem cells can be used to treat different conditions. These options include bone marrow transplantation, epigenetic reprogramming, producing islet cells to control hyperglycemia, using cells from SHED and using stem cells as alternatives to arthroplasties. There are several ongoing research projects that delve into human application of these treatments, with hope in the future to use them to cure chronic diseases. The next chapter will discuss the science involved with studying stem cells.

7

What is the Science Involved in Studying Stem Cells?

*Written By **Rishi Thangarajah***

INTRODUCTION

At their core, stem cells are unspecialized cells that possess the ability to differentiate into various types of cells, of increasing specialization. Due to this intrinsic property, stem cells are regarded as an important tool in the field of regenerative medicine, an excellent model for better understanding development biology and for studying the progression of diseases. The goal of this chapter is to provide you with an insight into the science and methodologies involved in studying stem cells. Specifically, the procedures used to determine aspects of directed stem cell differentiation, markers of pluripotency and factors that make stem cells an ideal model system will be discussed. The principles of the science behind studying stem cells will assist in supplementing the knowledge you have obtained so far. As mentioned in previous chapters, not all stem cells possess the same potential range for differentiation.

THE HIERARCHY OF STEM CELL POTENTIAL

There are many classes of stem cells that fall under this umbrella term. Totipotent cells stand at the top of the totem pole of potential cellular differentiation (Alison, Poulsom, Forbes & Wright, 2002). These cells can multiply and differentiate into all the cells contained within a human being, meaning that they possess the highest potential for differentiation (Zakrezewski, Dobryzński, Szymonowics & Rybak, 2019). A zygote, or fertilized egg, is an example of a totipotent cell because this cell acts as the progenitor for cells that will go on to form either the inner cell mass or the trophoblast (Zakrzewsk et al, 2019). As discussed in Chapter 5, four days after the zygote proliferates a structure known as the blastocyst is formed (Zakrzewski et al, 2019). The blastocyst consists of the aforementioned trophoblast, which forms extraembryonic structures like the placenta, and the inner cell mass (Alison et al, 2002). Embryonic stem cells are gathered from the inner cell mass (Zekrezewski et al, 2019).

Embryonic stem cells, or ESCs, are pluripotent cells that eventually proliferate and differentiate into clusters of cells known as germ layers (Alison et al., 2002). The three germ layers are the ectoderm, mesoderm and endoderm (Zekrezewski et al., 2019). These germ-layer cells are referred to as multipotent stem cells because they give rise to cells that will eventually differentiate to form the various types of tissue within the human body (Zekrezewski et al., 2019). Like all stem cells, multipotent stem cells can differentiate into more specialized cell types. However, unlike pluripotent stem cells, they can only differentiate into the cell types within one particular lineage (Alison et al., 2002). The lineage of a cell can be thought of as the family tree of a cell, denoting how a particular cell came to exist. An example of multipotent stem cells are the stem cells of the central nervous system that have the ability to differentiate into neurons, astrocytes and oligodendrocytes (Alison et al., 2002). At the bottom of the totem pole for potential differentiation is where unipotential stem cells sit (Alison et al., 2002). Unipotential stem cells are sometimes called committed progenitors because they are only capable of generating one specific type of cell (Alison et al., 2002). This is the case with most adult stem cells in the human body, for example stem cells in the basal layer of the epidermis only generate keratinized squamous cells (Alison & Islam, 2009).

USING STEM CELLS TO UNCOVER DETERMINANTS OF DIFFERENTIATION

Based on the information laid out in the subsection above, it could be interpreted that the lineage of every single cell in the human body can be traced back to a common pluripotent stem cell ancestor. This detail makes stem cells an appealing model system for researchers interested in studying the biological processes that underlie the development of multicellular organisms, such as human beings.

Think back to your adolescent years, and you may recall being told that *"you can be anything you want to be"*. However, as you progressed through life you naturally began to favour certain interests and activities over others. Consequently, by allowing yourself to specialize in certain aspects, you closed off the opportunity to specialize in others. Similarly, pluripotent stem cells differentiate into multipotent stem cells, which then differentiate into unipotential stem cells that generate cells specialized to carry out a specific function. Therefore, by undergoing subsequent levels of specialization, stem cells reduce their differentiation potential. Consider the factors that influenced your career choice. If you simplify it, the explanation behind your decision can be attributed to your surrounding environment, like the individuals you interact with, as well as your responses to these interactions. Maybe it was a science experiment done by your elementary school teacher that elicited your interest in science or, the fact that the people around you desired a typical day job, which fueled your pursuit of a career as an actress. Either way, in both cases an external stimulus elicited a response from you, which factored into your decision of what to specialize in.

Although cells cannot speak to one another in the traditional sense, they can communicate via a network of transcriptional factors to determine the fate of a pluripotent stem cell (Oh & Jang, 2019). By understanding how these transcriptional factors influence cell differentiation scientists can manipulate stem cells in order to generate the mature somatic cells they desire. The potential applications of these generated cells are discussed in greater detail within chapters 6 and 10. The following subsection will supplement the aforementioned topic by explaining the general science and methodologies behind research involving stem cells.

METHODOLOGIES USED FOR DIRECTING STEM CELL DIFFERENTIATION

Researchers interested in studying stem cells use a procedure known as cell culturing to carry out their experiments (Zakrzewski et al., 2019). Cell culturing is the process of growing cells outside of their natural environment, under the specific conditions laid out by the research team. Conditions that scientists must control include the pH level, temperature and specific conditions of the cell medium (Arora, 2013). The media of a cell culture usually consists of compounds involved in the regulation of the cell cycle, like amino acids or hormones, as well as an appropriate energy source to facilitate the growth of the cells (Arora, 2013). Therefore, by manipulating the media in which cells are cultured, researchers can ascertain how the extracellular environment influences stem cell differentiation (Zakrzewski et al., 2019). This information provides researchers with a better understanding of developmental biology. In addition to this, researchers can then manipulate cell culture mediums to create specialized cells that could be used for some of the treatments discussed in Chapter 6.

Researchers begin this procedure by acquiring human embryonic stem cells, or hESCs, from the inner cell mass of embryos that are in the early stage of their development (Zakrzewski et al., 2019). These pluripotent stem cells will then be transferred to a dish containing the selected medium using a procedure known as cell passaging (Zakrzewski et al., 2019). Since stem cells require a combination of growth factors and nutrients to undergo differentiation, the cell culture medium must be changed daily (Zakrzewski et al., 2019). This concept is similar to how one would remove a fish from its tank temporarily, in order to clean the tank. There are multiple procedures for cell passaging, but ultimately the procedure chosen by researchers will be dependent on the type of substrate used in the cell culture (Zakrzewski et al., 2019). It is also imperative that the investigators carrying out these experiments test each medium component at different concentrations and different points during embryonic development (Zakrzewski et al., 2019). The reason for this is because cell fates can differ when pluripotent cells receive differentiation factors of differing concentrations or at different stages in their development (Zakrzewski et al., 2019).

As mentioned in previous chapters, directed differentiation can yield many benefits in the realm of regenerative medicine. For instance in 2009, Idelson and her colleagues published a study on the direct differentiations of hESC into retinal pigment epithelium, or RPE, cells. The loss of this epithelium is caused by a disease caused by ageing, called retinitis pigmentosa, and it is the leading cause of visual disability in the western hemisphere (Idelson et al., 2009). Although treatments are available, the reality is that many patients with retinitis pigmentosa end up losing their vision (Idelson et al., 2009). Due to the nature of this disease, the application of stem cells to regenerate lost RPE cells appears to be a viable method of slowing down the progression of the disease (Idelson et al., 2009). Through the implementation of defined conditions for the cell culture, Idelson and her colleagues were able to show that nicotinamide and Activin A can lead to the differentiation of hESCs to RPEs (Idelson et al., 2009). When nicotinamide interacted with hESCs, it promoted neural cell differentiation, and eventually RPE differentiation (Idelson et al., 2009). The results of this study exhibit that cell differentiation can be directed. This example reiterates how beneficial stem cell research can be for the field of regenerative medicine. Although this is not the only type of stem cell research that is conducted.

METHODOLOGIES USED FOR IDENTIFYING STEM CELL PLURIPOTENCY

In the previous subsection, it was stated that the extracellular environment plays a role in determining cell fate. However, neighbouring stem cells also play a pivotal role in determining the cell fate of the pluripotent cells around them (Chen, Fitzgerald, Zimmerberg, Kleinman & Margolis, 2007). Stem cells, like all cells, contain genetic information known as DNA. Genes, the basic unit of hereditary information, are formed by segments of DNA. Genes contained in a cell can be expressed to create traits which are expressed in a person.. Researchers studying stem cells refer to the genes and their subsequent products associated with pluripotency as stem cell markers.

To identify what genes and products are involved in determining stem cell pluripotency, researchers employ a procedure known as a reverse transcriptase-polymerase chain reaction, commonly referred to as RT-PCR (Rhee & Bao, 2009). Reverse transcriptase is an enzyme that creates corresponding DNA from a DNA template (Huggett, Dheda, Bustin & Zumla, 2005). PCR is a sophisticated process that amplifies the quantity of DNA, allowing scientists to study the DNA sequence from the samples (Huggett et al., 2005). PCR can be thought of as a DNA photocopying process. First, the DNA sample must be denatured to separate the two DNA strands (Huggett et al, 2005). Then a primer used to select the segment of DNA amplification is annealed to the region of interest (Huggett et al., 2005). A DNA polymerase called Taq facilitates the addition of free nucleotides to synthesize copies of the original DNA strands (Huggett et al., 2005). After this process is complete, a new double DNA strand is created from the original. In a study that used this technique, the research team found that the differentiation of pluripotent stem cells was marked by a decrease in OCT-4 mRNA expression (Rhee & Bao, 2009).

The pluripotency of undifferentiated embryonic stem cells can also be determined by identifying the presence of certain cell surface and molecular markers (Zhao, Ji, Zhang, Li & Ma, 2012). For this identification process, researchers utilize a procedure called flow cytometry. Put simply, flow cytometry detects cells of interest, on an individual basis, as they "flow" towards a laser (Jahan-Tigh, Ryan, Obermoser & Schwarzenberger, 2012). The preparation process involves the centrifugation of the sample of cells, followed by the addition of fluorescent antibodies of the cellular component of interest (Jahan-Tigh et al., 2012).

Cell surface antigens are one of those cellular components of interest that can be fluorescently tagged. Antigens are molecular structures on cells that can be interacted with to elicit cellular responses. Fluorescent tagging allows scientists to visualize the presence of specific cellular components, like antigens. Stage-specific embryonic antigens, or SSEA, are involved in dictating cell surface interactions during embryonic development (Zhao et al., 2012). A cluster of differentiation, or CD, antigens are another group of surface proteins expressed by stem cells (Zhao et al., 2012). These antigens are shared by cells of the same class, which makes it seamless for researchers to quantify cell cultures by cell class using flow cytometry (Zhao et al., 2012). Jean-Marie Ramirez and his team used flow cytometry to determine which markers most indicative of pluripotency out of eight widely known ones (Ramirez et al., 2011). Based on the results from this procedure, they suggested that the presence of OCT-4, SSEA3 and TRA-1-60 were more indicative of pluripotency than other known markers like CD24 and NANOG proteins (Ramirez et al., 2011). The findings of this research indicates that the disappearance of these markers of pluripotency initiates differentiation.

THE STUDY OF INDUCED PLURIPOTENT STEM CELLS

So, if the disappearance of pluripotent markers initiates differentiation, then it stands to reason that the reintroduction of those same markers could induce pluripotency after differentiation. In 2006, Kazutoshi Takahashi and Shinya Yamanaka proved this by demonstrating that differentiated somatic cells could be reverted to their pluripotent state (Takahashi & Yamanaka, 2006). They accomplished this feat by introducing 24 candidate genes into the cell culture of adult mouse fibroblasts and determining which ones induced a pluripotent state upon the cells in the culture (Takahashi & Yamanaka, 2006). They conducted an assay analysis to carry out this study, using G418 resistance as a method of assessing the pluripotency of the cell (Takahashi & Yamanaka, 2006). A biological assay is an analysis tool used by scientists to test the effects of a compound on cells or tissue samples. G418 is an antibiotic that results in the cell death of cells without the neomycin resistance gene (Tanabe, Takahashi & Yamanaka, 2014). This resistance gene happens to be expressed in embryonic stem cells but not in somatic cells, which made G418 an effective diagnostic tool for assessing pluripotency (Tanabe, Takahashi & Yamanaka, 2014). Based on the results of this experiment, Takahashi and Yamanaka determined that introducing OCT-3, OCT-4, Sox2, c-Myc and Klf4 to mature fibroblast cells, under embryonic stem cell culture conditions, resulted in their reversion back to a pluripotent cell state (Takahashi & Yamanaka, 2006). Future therapeutic applications and research involving induced pluripotent stem cells are explored in Chapters 6 and 11 respectively.

USING STEM CELLS AS A MODEL SYSTEM

In the biological sense, a model is an organism utilized by researchers to assess a biological phenomenon from another organism. You may have noticed that preclinical trials utilize an animal model, like rodents, to test the effectiveness of a drug. The reason these models can be used instead of humans is because they are homologous, meaning they share many similarities to humans in their genetic profile and physiological systems (DeBry & Seldin, 1996). However, the results that occur when using animal models do not always translate to humans. In addition to that, the ability to differentiate into multiple types of cells makes stem cells the perfect model system to uncover unknown disease progression and test drug effectiveness specifically for humans (Sterneckert, Reinhardt & Schöler, 2014).

In a study that examined the relationship between chronic lymphocytic leukemia or CLL, B-cells and bone marrow stromal cells, mesenchymal stem cells were used as a model. Previous studies showed that CLL B-cells cannot survive without interacting with bone marrow cells (Ding et al, 2009). To better understand why this was the case, mesenchymal stem cells were grown alongside CLL B-cells to mimic the parameters between a CLL B-cell and bone marrow stromal cell (Ding et al, 2009). The researchers co-cultured the cells and used flow cytometry to determine expression factors that influenced disease progression (Ding et al., 2009). They found that the upregulation of CD71, CD25, CD69, CD70 and CD71, especially the latter, were upregulated within the co-culture (Ding et al., 2009). They were also able to determine that this activation was bidirectional and most likely mediated by soluble factors of both cells (Ding et al., 2009). In a similar way to this experiment, the effectiveness of potential drugs could be tested on cultures of stem cells affected by diseases. Researchers could use assays to determine whether the stem cell is disease-free after the intervention, and also observe any harmful side effects caused by the treatment. The potential of what researchers can do with stem cells is examined in extensive detail within Chapter 10.

CONCLUSION

This chapter went over aspects of the science involved in studying stem cells. Stem cells are an umbrella term that contain classes of different cells distinguished by their potential to differentiate. The cell fate of pluripotent cells are determined by innate gene expression, neighboring cells, and the extracellular microenvironment. Stem cell differentiation can be directed by researchers by manipulating the media on which stem cells are cultured on. The pluripotency of a cell can be determined by identifying stem cell markers through the use of RT-PCR and flow cytometry. These stem cell markers can be reintroduced to somatic cells to revert them to a pluripotent state. Finally, stem cells can be utilized as models to better understand the progression of diseases and the outcomes of pharmaceutical treatments. The next chapter examines the ethical dilemmas posed by stem cells research.

8

What Controversy is there Surrounding Stem Cells?

*Written By **Anusha Mappanasingam***

IN CHAPTERS **2** AND **3** RESPECTIVELY, we discussed the discovery of stem cells and the profound impact they had on society. In the previous chapter, the science involving stem cells were presented. From the information discussed, the high potential that stem cells hold for the present and future becomes obvious. While future research can ensure our knowledge on stem cells is always growing, it is the impact that stem cells can have in improving quality of life that is the most astonishing. One of the major ways that quality of life is changed is through the potential treatments that stem cells provide. In the wake of the overwhelming impact of stem cell research, individuals feel gratitude for the life-altering treatments offered with stem cell research. Despite the advantages it provides for therapeutic purposes, the usage of stem cells and stem cell research have sparked controversy throughout the world. In this chapter, the ethical dilemmas behind the controversy of embryonic stem cells and human cloning will be presented, along with some potential alternatives to cope with the specific controversies.

With the accompanying rise of technology in the digital age, controversies are common: with the in-depth knowledge that we have now obtained, individuals have gained a deeper understanding into worldly issues, allowing for multiple complex viewpoints. These viewpoints often clash and can result in heated debates and policy changes. While the controversies can stem from political arguments or the uncovering of facades worn by celebrities, it is usually ethical dilemmas that cause these controversies to become so popular. This was, and still is, the case with stem cell research.

Specifically, it is the use of embryonic stem cells that causes the most outrage. Consider the idea that humankind made a discovery that was so promising that with proper research, new therapies could be developed to save lives. This is something that most would rejoice in and this discovery would be welcomed wholeheartedly. But then, what if it was found that this discovery could also cost lives? It is a similar idea that has made the use of embryonic stem cells controversial. Despite their evident potential in research and therapy, their origins and the process of obtaining them presents a problem. As discussed in chapter 5, embryonic stem cells are obtained right from the embryo. When researchers extract these stem cells, the embryo that they come from is destroyed in the process (Holm, 2002). When referring to the source of these stem cells as an "embryo", it is easy to disregard its demise; especially for the general public. The science nature of the term can allow for a disassociation that prohibits us from understanding what the embryo really is. Thus, the end of an embryo can be of minimal significance. This is why it is important to remind ourselves of what is really meant by the term: an embryo is the entity that exists at an early stage of development and is formed through the sperm fertilizing a cell in the ovary (Findlay et al., 2007). When taking into consideration that an embryo is a form that we once held, its destruction might now have more significance to us: how can we allow someone to destroy an undeniable component of human life so willingly? While this is a viewpoint that many carry, there is also the opposing side to consider: many believe that embryos are only a group of cells, and thus do not require the considerations and rights we would give a fully-grown human (Holm, 2002). Ultimately, this is where the controversy comes from: while for some, the idea of destroying embryos becomes unfathomable once its association to human life is taken into account, others are fiercely determined that using the embryo is not an issue of ethical concern due to its lack of human resemblance. When considering the former opinion mentioned, this destruction of human embryos causes us to deliberate the potential suppression of human rights (Holm, 2002). It is well-known that as human beings, we are entitled to our right to live; this is a basic human right that all are entitled to. When considering this human right, the opposing viewpoint to the usage of embryonic stem cells finds its fuel: if humans have the right to live, why not human embryos? If individuals can agree on an embryo being the initial form of human development, then what happens when the right to exist and grow is stripped away with the embryo's destruction? In order to begin considering the potential answers to these questions, the implication of these questions need to be addressed. By considering these questions to be valid, the idea that human life begins at conception is implied (Brock, 2006). If one considers the contrary to what has been implied—that society was able to come to the conclusion that human life does not begin at conception but rather begins once the child is born—then the embryo's destruction would not be a cause for concern. However, both of these viewpoints are not the case; despite the decades of discussion on the topic, the controversy still persists. While this claim will be debated and considered later on in this chapter, it is still important to highlight that this truly is the fundamental question proposed by the controversy surrounding stem cell research. For argument's sake, let us accept the idea that human life begins at conception to be a well-established fact. This would mean that a human embryo is entitled to human rights. But then, to what extent? That is, if one was presented with a scale indicating the vigority with which human rights were exercised, then where would human embryos fall with respect to fully-grown humans? The answer to this question is crucial, as it would ultimately determine whether the tradeoff of an embryo for saving a grown human is acceptable; if the embryo is lower on the scale with respect to ourselves, then this tradeoff might be tolerated (Brock, 2006). However, if the contrary is true—if the embryo is at the same level as ourselves on the scale of human rights—researchers and advocates for stem cell research will struggle with justifying the use of the former (Brock, 2006). Again, it is important to recognize what individuals are essentially deliberating when engaging in this controversy; society is questioning the ethical limits we must implement in order to dignify the life of embryos on the chance that their status is agreed to be equal to ourselves (Holm, 2002).

While most continue to encourage this debate in hopes of finding a possible middle-ground that allows for the usage of embryonic stem cells while taking into account the ethical demands, there are many others that cannot fathom the use of embryonic stem cells. These are usually religious groups that are behind the movement to ban the use of embryonic stem cells (Farley, 2009). Specifically, the idea that the use of embryonic stem cells for research and therapeutic purposes is an unethical practice is one held by many Roman Catholics (Charitos et al., 2021). The primary argument for their stance reflects what has been discussed thus far: human life begins at conception, and so by destroying an embryo, one is ultimately ending human life and this goes against the very core of Roman Catholic beliefs (Charitos et al., 2021). Similar viewpoints are echoed by other religions, such as the Orthodox church and Hinduism, although these viewpoints are more particular than the one belonging to Roman Catholics (Charitos et al., 2021). More details on these religious viewpoints will be accounted for in chapter 9. Additionally, religion is not the only source for the debate around stem cell research and therapies with embryonic stem cells; this controversy has a political impact as well (Nisbet et al., 2003). This was primarily seen through the inconsistencies seen in the US with the Bush administration and their stance on stem cell research using embryos (Nisbet et al., 2003). When considering the conflict that arose within the Bush administration—with some wanting to support stem cell funding and others wanting to ban it in order to please Catholic supporters—individuals are confronted with the controversy's political impact (Nisbet et al., 2003). While more details regarding the stem cell policy changes that occured with the change in leadership from Bush to Obama will be discussed in chapter 9, it is important to recognize the complexities that are associated with the stem cell research: this is a controversy that society will struggle to find common ground for (Murugan, 2009). We will see this shortly in the chapter, but solutions to this debate will never be simple and will have implications in all areas of a person's life.

While the implications of this controversy are grand and complex, there are solutions suggested to tackle the alleged unethical nature of embryonic stem cells. These solutions stem from one general concept that appears to be simple and effective at first glance: researchers have suggested using alternative sources to embryos for stem cell research and therapy (Brock, 2006). Let us consider some of these possible solutions and clarify the extent to which they make worthy replacements of embryonic stem cells. As discussed in chapter 5, there are many different types of stem cells that can be used in research and therapy. Most of these types of stem cells appear to be potential alternatives to embryonic stem cells. One source that could replace embryos is adult stem cells (Brock, 2006). Using this source would eliminate these ethical issues since the use of adult stem cells does not involve the annihilation of any human form; this would mean that the deliberation of moral status of human embryos would be unnecessary (Brock, 2006). While this eliminates most of the original concern sparked by the debate discussed in this chapter, the usage of adult stem cells introduces some of its own concerns (Brock, 2006). Adult stem cells are believed to be inferior to embryonic stem cells (Brock, 2006). They are considered as such due to their limited ability to differentiate into the desired cells (4. The Adult Stem Cell | Stemcells.Nih.Gov, n.d.). This coincides with having a limited therapeutic ability to save lives (Brock, 2006). Another alternative source of stem cells that could be used is those located in umbilical cords, also referred to as cord blood cells (Brock, 2006). In contrast with adult stem cells, umbilical cord stem cells appear to be readily available and have a higher ability to differentiate into other cells (Weiss & Troyer, 2006). An additional benefit is their inexpensive nature: the umbilical cord is typically disposed of immediately after birth, and therefore is easy to obtain (Weiss & Troyer, 2006). It is important to note that the substitute sources discussed thus far are limited in ethical controversy. The following sources that will now be considered have serious ethical considerations that are both similar to and serve as additions to what has been discussed thus far.

Generally, aborted fetuses and unfertilized eggs are recommended as alternative sources due to their potential to otherwise be wasted and discarded (Brock, 2006). When considering the fact that the refusal to use both sources would result in them being discarded, using them appears to be a resourceful option. However, one must consider the heavy ethical dilemmas surrounding them both. Similar to embryonic stem cells, the idea of using an embryo brings controversy, even though these sources are considered to be less problematic (Holm, 2002). Something of popular debate is whether the use of these sources are truly justified: is it better to use something so that it is not needlessly killed, or is conservation merely an unworthy excuse? While this is unavoidably up for debate, the real concern that accompanies these sources is when confronting the individuals providing either the aborted fetus or unfertilized eggs for therapy and research purposes. How does one ask an individual who has just given up their child or has experienced a failed IVF procedure to give up these parts of themselves to researchers in a lab? Even if the benefits are explained, this can cause a great amount of emotional distress and places these individuals in vulnerable positions. Extreme sensitivity would be required to tackle such situations, which all researchers may not be qualified to provide. Overall, while considering all of these alternative sources' ability to be equivalently effective as embryonic stem cells as questionable. Continuing research into this area is encouraged due to its relative lack of ethical considerations when compared to embryonic stem cells.

It is important to recognize that while the moral status of embryonic stem cells accounts for most of the controversy surrounding stem cell research, it is not the only idea that jeopardizes the future potential of this research. Again, we return to reflecting on the source of these embryonic stem cells. While thus far we have only pondered on the ethical dilemmas presented by using the embryo, let us now consider the dilemmas presented while obtaining the embryo. When obtaining the embryos, the required ova to get these stem cells must come from women (Holm, 2002). This raises concern over the coercion and exploitation of women that could occur as a result of this process (Holm, 2002). Women in desperate situations could be exploited, and thus forced to give up their ova; these situations would most likely include an unfair power dynamic in which those experiencing great financial distress are exploited. It is important to highlight that this exploitation does not have to occur deliberately to be unethical: individuals who are in desperate situations—regardless of the reason—would still be violated, even if no one is seemingly forcing them. With this in mind, the lines between what is coerced and willing can become blurred, and thus not only cause harm for women in the moment of donation, but it can also cause legal repercussions as well. If individuals were to follow through with obtaining embryos in this manner, careful rules and training would have to be implemented to ensure the safety of donors, and eliminate the possibility of coercion as much as possible. An alternative solution is using spare embryos as a source (Holm, 2002). These would be slightly limited in the coercion that could occur since the embryos would be being discarded regardless. However, as discussed previously, the usage of spare embryos still presents its own set of ethical concerns that require its own solutions. In one paper, the usage of bovines—a type of animal—with nuclear replacement techniques is suggested (Holm, 2002). This could be a worthy solution if society concluded that human embryos were of higher moral status than other embryos (Holm, 2002). However since this is not the case, coming up with this conclusion would initiate the ethical debate of the value for human life vs. animal life (Holm, 2002). This is an ethical dilemma that, like others mentioned, has very few resolutions.

Another source of controversy that arises from the use of embyronic stem cells is human cloning (Jaenisch, 2004). Before we begin discussing this controversy, let us go over some details of human cloning. Cloning is the process performed to obtain a genetically identical copy of a biological entity (Häyry, 2018). Cloning typically occurs through the process called somatic cell nuclear transplant (SCNT) (Gouveia et al., 2020). As discussed in detail in chapter 6, SCNT is the process in which the nucleus of an egg is replaced with the nucleus of a mature adult cell. When SCNT was originally discovered with frog cells in the mid to late twentieth century, its discovery allowed for the possibility of both reproductive and therapeutic cloning (Briggs & King, 1952; Gouveia et al., 2020).

When considering the controversy surrounding human cloning, it is important to recognize that it is firstly the debate around the status of a human embryo that makes human cloning unfathomable (Brock, 2006). As discussed previously, questions determining the point at which human life begins and the extent to which one form is more valuable than others need to be answered in order to arrive at a form of resolution regarding this specific source of controversy (Brock, 2006). A potential solution to this has been found by using induced pluripotent cells (iPSC) (Tan et al., 2016). When iPSCs eliminate the use of embryonic stem cells by replacing SCNT, they prove themselves as worthy solutions to explore (Tan et al., 2016). However, their therapeutic effectiveness is debatable due to their use of stem cells from sources excluding the embryo (Tan et al., 2016). More details on iPSCs are discussed in chapter 6.

While it can be agreed that without addressing this ethical debate researchers cannot move forward with human cloning, it is important to highlight that the idea of human cloning itself creates controversy. If the argument regarding the moral status of the human embryo is resolved and common ground is reached, there will be few objections for therapeutic cloning (Häyry, 2018). However, this is not the case for reproductive cloning (Häyry, 2018). If permitted, reproductive cloning would be performed for reproductive purposes, including couples experiencing infertility problems (Sciences (US) et al., 2002). The problems that arise from this have to do with the idea of the extent to which humans should be allowed to control natural selection and the course of humanity (Häyry, 2018). Aside from the fact making human babies in an asexual manner is unnatural and can be slightly disturbing to some, individuals need to carefully consider what the future would look like were this allowed: how would society implement regulations with respect to reproductive cloning? If reproductive cloning was implemented, what is the guarantee that individuals would not take it too far, and begin to seek these processes in order to design the perfect child (Häyry, 2018)? Many advocate against human cloning as they have begun to understand that it is our humanity that is at stake in such a scenario (Häyry, 2018). While this might seem trivial to some, the implications of this on human interactions and daily life are immense and require thoughtful consideration before proceeding any further.

In conclusion, this chapter presented and evaluated the controversy surrounding stem cell research. Firstly, the primary ethical concerns regarding the use of embryonic stem cells was considered where the moral status of the human embryo comes into play. Religious and political views regarding the controversy were briefly mentioned. Potential solutions were offered, including finding alternative sources to replace embryos. But these solutions were found to carry their own ethical and scientific problems. Individuals were then presented with another source of controversy: women who donate ova are at risk for coercion and exploitation. Stricter protocols could be implemented to tackle this, or one could use alternative sources rather than women, but these come with their own legal and ethical repercussions. Lastly, the controversy with human cloning was discussed: aside from the moral status of the human embryo in question, there are growing concerns with what engaging in human cloning could do to our humanity.

Due to the prevalence of these controversies regarding stem cell research, members of society are often found eager to contribute their opinions regarding this topic. In the next chapter, the portrayal of stem cell research in popular media will be discussed.

9

How is Stem Cell Research Portrayed in Popular Media?

*Written By **Rodala Aranya***

INTRODUCTION

The primary line of communication between science and the general public is the media. Public attitudes towards innovations, scientific findings, and research can be swayed by what we consume in media. Whether it's shedding a positive light on discovery or raining hail on its negative possibilities, the dialogue that the media produces shapes public opinion and decisions of policy-makers and governments regarding that research. This rings especially true when we examine the biomedical field of stem cell research.

Stem cell research has been a hot topic of public debate since the late 1990s. It seems, even though this is not the case, that the general narrative around stem cells in the media is one of the varying extremes. Either stem cell research is a fantastic breakthrough for treatment or a massive disaster regarding our ethics, morals and future of humanity. These are not the only two dialogues in public, but still, the media tends to sensationalize news stories.

When looking to analyze the media's portrayals of stem cell research, it can be seen that recent discourse surrounding the topic has become stagnant, maybe due to no breakthroughs in the last few years, so for the purposes of the chapter, we will examine stem cell narratives that have dominated our media in the last twenty years. First, we will examine the controversy surrounding the use of embryonic stem cells, which was discussed in more detail in chapter eight, and how it was portrayed in the media. Then, we will explore another interesting avenue of discussion that arises when critiquing stem cell research, which is gene editing. The rest of the chapter will focus on the shift in media portrayal of stem cell research from concern to the rise of the "stem cell hype.". Lastly, we will evaluate how this perception has recently been utilized to promote fraudulent therapy clinics.

STEM CELL APPREHENSION: THE ETHICAL DEBATE OF EMBRYONIC STEM CELLS

Since the early days of their discovery, stem cell research has been met with excitement for its future possibilities, but it has also faced its brunt of social controversy. One of the most well-known discourses surrounding stem cells in the media was surrounding the usage of human embryonic stem cells (hESC). As explored in chapter eight, this was one of the most significant controversies surrounding stem cells from the early 2000s to the 2010s. This interest in stem cell research can be seen throughout various domains—mass media, the press, academic settings, policy forums, legal proceedings and more (Zarzeczny & Caulfield, 2009)"In 1994, fewer than one hundred articles were published on the topic of stem cells, but by 1999, this output more than doubled to more than two hundred articles published. By 2001, research had more than tripled over 1994 levels, with more than three hundred articles appearing" (Nisbet et al., 2003, pg 51).

The problem arose because earlier stem cell therapies solely relied on embryonic cells. The main underlying question that captured the public was whether, and to what degree, human embryos have a moral status requiring protection (Holm, 2002). This section will not go into depth concerning the controversy but instead just look at how the portrayals played out. The ethical debate is covered in more detail in the previous chapter.

For quite a while, stem cell research was enshrined in this controversy. The nature of the representations made in the media surrounding stem cell research during this period is very important to examine, as both informed and influenced the public. The topic became one of a policy issue as well. Let's take a look at how the controversy came to be.

In 1998, the first human embryonic stem cells were isolated and grown in the lab. Leading up to 2001, public discourse was fraught with the implications of what this meant. As thoroughly explored in the previous chapter, many religious and pro-life groups saw the use of embryonic stem cells to be killing a human life in the process of research and demanded that it be banned. It became a matter of public opinion, where the moral status of human embryos and ethical limits concerning them came to the forefront. News outlets framed this narrative as black and white, as science fighting with religion (Nisbet et al., 2003).

In 2001, President Bush limited federal funding towards research on human embryonic stem cells but allowed research to continue for existing embryonic stem cell lines. Following the election of President Obama, the 2001restrictions were lifted. Media reporting on stem cells heavily relied on dramatic storytelling (Zarzeczny & Caulfield, 2009). As politics came into the conversation, proponents and opponents of stem cell research employed intense narratives of "playing God" or "being irrational religious zealots opposed to progress."(Nisbet et al., 2003, pg 50). Suffice to say, these representations only sensationalized stem cell research. During this time, media discussion around stem cell research focused heavily on hESC research and focused very little on other promises of stem cells (Dresser, 2010).

In a paper published in 2009, titled "Emerging Ethical, Legal and Social

Issues Associated with Stem Cell Research & and the Current Role of the Moral Status of the Embryo" researchers surveyed various news sources, such as print media determine how the moral context of embryonic cells played out in the general public. Rather interestingly, it was found that popular media did not promote any one belief but focused on identifying the belief of others, such as President Bush, the Roman Catholic Church or scientists, and centred around the ethical debate the country was engaged in. So, for the years following the Bush administration into Obama's first term, stem cell research discourse heavily centred around the morality of using embryonic cells (Zarzeczny & Caulfield, 2009).

From the beginning of Obama's terms in office, media portrayal saw a dramatic shift in coverage. News stories and consequently public attention shifted to clinical translation, focusing on the research, discovery and development of treatments using stem cells. This may be due to the reversal of the controversial and unpopular ban on research funding for hESC and the public's evolving perceptions regarding the nature of the stem cell research field. Before we look into this new optimistic shift in discourse, there is one more perspective to look at (Kamenova, 2017).

STEM CELL APPREHENSION: FUTURE OF HUMANITY AND DESIGNER BABIES

Another big aspect of stem cell portrayal in the media has strayed away from research into the implications stem cell breakthroughs could have for our future. Concerns of human cloning and designer babies arise when discussing stem cells, and it is a point of negativity surrounding the portrayal of stem cell research (Hynes et al., 2017). So, what is being said, and what truth does it hold? We could explore many discussion points, but for the purpose of this chapter, we will look primarily at the conversation surrounding designer babies. For information regarding the controversy of human cloning, refer to chapter eight.

Designer babies have been reported in the news since the first "test-tube baby" was born in 1978. The phrase in itself is not accurate. Despite the name, "test-tube babies" are not developed in a test tube. There are narratives of a dystopian future where racks of tailor-made babies will be waiting in their "numbered test tubes." In 2016 Author Kazuo Ishiguro, who wrote the novel Never Let me Go, warned that with gene editing, "we're coming close to the point where we can, objectively in some sense, create people who are superior to others." This is of worry to many people due to the implications of how society may function if we go down this path. Concerns regarding superior babies, future caste systems, elimination of many disorders that people of those communities identify with, and the creation of designer baby armies are just some of them (Hynes et al., 2017).

The uproar around modified babies was heightened with the gene-editing method CRISPR-Cas9, which uses natural enzymes to target and snip genes with pinpoint accuracy. But designer babies have not come and are nowhere close to reality, not in the way the public views it. The medical research community itself is not an advocate for the idea of using Crispr-Cas9 for human reproduction (Hynes et al., 2017). Many scientists have also pointed out that gene editing would be hard and expensive, with unknown health risks. The regulatory landscape is very unfavourable for creating any sort of super baby (Verlinsky, 2005).

Even if the concept of gene editing comes to the forefront, which is unavoidable, it will not be until far in the future, as stem cell research is focused primarily on stem cell therapies today. Stem cell research is distanced in many senses from gene sequencing, which in the future may allow people to produce a huge number of embryos and read their genomes. One of the biggest issues is that the media often conflates the two, and this is where we see misinformation spread. Possibly, stem cells may have a role to play in the future where people may grow eggs from stem cells to have an abundance of eggs to choose from, but even then, it would not be the stem cells themselves that lead to these designer babies (Hynes et al., 2017).

But as mentioned previously, the "selection" of certain traits or genes is not as easy as clicking an option on the screen. Biology is complex. Human traits are complex. Characteristics associated with designer babies – height, intelligence, physical abilities – are not controlled by just one or a few genes. Another point of interest regarding designer babies is that we may be able to modify genes to exclude certain diseases. Research is being conducted, and future debates may ensue regarding what diseases should be prevented by editing. So, yes, in the future, there is a possibility of gene editing that may have to do with stem cells, but society is far from that outcome (Verlinsky, 2005). This does not stop our discourse from being overtaken with concepts such as these when discussing stem cell research, which inevitably hinders true dialogue from taking place.

STEM CELL HYPE: UNWARRANTED OPTIMISM

In the last decade or so, stem cells have been primarily used in clinical trials for treatments for cancer and bone marrow diseases such as graft-versus-host disease. Despite this, media coverage seems fixated on the impact of stem cells on diabetes, cardiovascular diseases and neurological conditions. Research has shown that the media and the general public seem to misrepresent the current advances in stem cell therapy. Despite some of the controversies explored in the previous chapter, public opinion and expectation towards stem cell therapies have remained quite high. Looking back at opinion polls that were conducted in 2015, results showed that "expectations about benefits associated with SC therapies outweigh perceptions of risk" (Kamenova & Caulfield, 2015, pg 2).

Overall, the portrayal of stem cell research in the media has recently taken an overly optimistic slant, especially when concerning stem cell therapies. Headlines have focused on breakthroughs being just around the corner. This is supported by a correlated increase in public expectations and a newfound optimism for biomedicine in the last decade. 69 percent of news reports showed that the media has been overestimating stem cell therapy timelines, pushing the narrative that many will be available within the next five to ten years. Let's keep in mind that the study which reported this was conducted in 2015, looking at articles from 2010-2013, so we have already reached "the corner" with relatively few breakthroughs in terms of therapies that can be conducted on humans (Kamenova & Caulfield, 2015). More recent studies on media coverage around stem cells have shown that this perception has remained.

It is really important to note that despite the excitement, the timeline for most stem cell therapies is in their infancy stages. The current approved list of stem cell therapies in the US and Canada is less than a dozen combined (Caulfield & Murdoch, 2019). This raises the question of how is this dialogue shaping public perception and what does foster these unexpected timelines contribute to? We see this play out in society. As hope for stem cell therapies rise, certain companies are known to take advantage.

STEM CELL HYPE: DISHONEST MARKETING OF THERAPIES AND PRODUCTS

As this newfound positive perspective and media representation of stem cells has taken off in the last few years, companies and people have taken advantage of the new discourse. Companies are taking these optimistic narratives and misconstruing research, marketing correctly to a specific audience. Stem cells have been coined as anything from the "cure-all" to "a miracle treatment." Through taking advantage of this buzz, dishonest providers have been able to sell stem cell products and treatments that are both unproven by science and unapproved by the FDA in the US and Health Canada. It has gotten so bad that health authorities have had to put out public statements warning people of these illegal scams (Caulfield & Murdoch, 2019).

While it has become routine to use stem cells from bone marrow or blood transplant procedures to treat cancer and disorders of the blood and immune system, there are only a few proven and approved uses of stem cells. This point of interest seems to get fairly misconstrued in the media. If you were to look up stem cell therapies and a location, your computer screen would light up with hundreds, if not thousands, of results. Many of them are reputable sources talking about approved therapies, but if you were to get specific on a search, advertisements, products and therapies pop up from various countries (Marcon et al., 2017).

So, who is the target audience here? Unfortunately, many of these clinics offer new hope to patients who have run out of other options. Even though they are experimental and may cause potential harm, treatments can seem to be better than the alternative—doing nothing. But the issue here is scientifically backed treatments have become harder and harder to separate from dangerous ones. As stem cell research discourse fosters optimism and hope, the market has taken its own shape. On the rise are fringe clinics promising to cure anything from Alzheimer's disease to blindness (Murdoch et al., 2018).

Clinics will promise big claims to people looking for other alternatives. Anecdotal testimonials online, in the form of written statements and videos, will be posted on websites and social media accounts to try to provide the validity of their products through "real human experiences." Some clinics have even gone as far as to post their work to imitate other reputable scientific works on ClinicalTrials, a US federal repository that cannot always vet postings to make sure the work meets federal standards (FDA, 2020).

Sean Morrison is the director of Children's Medical Research Centre at UT Southwestern and the chair of the International Society for Stem Cell Research public policy committee. He stated, "It's completely impossible for individual patients, even individual physicians, to evaluate the evidence underlying every therapy that's out there and to make an informed decision" (FDA, 2020). This wild west of misinformation and misleading practices and products has forced many in the field to bring to attention this issue and correct the record, such as the Mayo Clinic and the California Institute for Regenerative Medicine in the United States (Murdoch et al., 2018).

What a lot of these clinics have in common is the promotion of a "one size fits all" treatment. From Alzheimer's to knee pain, it seems as if all the treatments are there. Some clinics were removing a patient's fat cells and then injecting them back in, with the claim that they were stem cells! One of the biggest concerns that are voiced is the danger of these stem cell clinics. Not only are they not FDA approved, costs thousands of dollars and promises false hope to many desperate people, but the treatments or products from these clinics can also actually be dangerous for one's health. There have been cases of patients being blinded or developing tumours after these therapies. The FDA in the US and Health Canada have been cracking down on many of these clinics and have increased efforts to regulate the industry more rigorously in the last few years. Many clinics have been shut down, and several have been sued for deceptive advertising of stem cell therapies (Caulfield & Murdoch, 2019).

An FDA spokesperson explained, "We are concerned that people may use these products with a false sense of security about their safety and efficacy. As such, the agency has stepped up its oversight of cellular and related products in recent years and has issued compliance actions, including numerous warning and untitled letters, and pursued litiga- tion for serious violations of the law, including some involving patient harm" (FDA, 2020, para 4).

CONCLUSION

Stem cell research in the public seems to be quite a matter of contention. Media discourse has created a few powerful narratives in which we see stem cells benign discussed.

On one side, stem cell apprehension questions the use of hESC and impli- cations of gene editing. On another, stem cell hype fosters unrealistic timelines that are then misconstrued to promote dishonest services. Either it is a concern of ethics and humanity's future, or it is the ground-breaking answer to many of our problems. To be clear, these are not the only discussions that are had around stem cell research, but through a thorough examination, it is evident that these are some of the leading talking points.

So, where does this leave us? Even though the portrayal of stem cells has been divided in the past, studies show that the public is taking a more positive outlook on this scientific endeavour. At the very least, people are intrigued by science and its potential to revolutionize the medical field. We must remember that stem cell research is still in its infancy stages, and therapies and treatments will come, but we must be patient. Concerning ethical dilemmas we may face in the future, it is crucial to have these conversations so that society can determine where we stand. Still, we must be careful not to get too ahead of ourselves so as to hinder any positive progress we may make. Currently, there is no basis for any concerns regarding stem cell and human editing; we are far from it. To stay grounded in the scientific truth, we as the public must strive to stay educated on these matters and listen to the experts. Instead of running with the most dramatic headline, our opinions should be based on facts and realistic timelines. This brings into question, what truly is the future of stem cells? The concluding chapter will focus on just that.

10 ● What Future Direction Will Stem Cell Research Take?

*Written By **Joonsoo Sean Lyeo***

INTRODUCTION

When embryonic stem cells were first isolated from mice in the 1980s, few could have predicted the tremendous amount of intrigue they would eventually garner (Solter, 2006). In the past few decades, stem cell research, and its alleged potential to revolutionize the field of medicine, has received an unprecedented level of public interest (Bongso & Richards, 2004). What had once been considered a highly-specialized and relatively niche subject, only of interest to those in the relevant academic circles, now evokes great expectations for the future of healthcare (Solter, 2006). That being said, it should be acknowledged that stem cell research is still in its infancy, and there are still significant obstacles to overcome before its full potential can be ascertained (Zakrzewski et al., 2019). These obstacles, and the concerted efforts to overcome them, are likely to guide the future direction of stem cell research.

ANIMAL TESTING

For starters, as discussed in Chapter 9: What Controversy is There Surrounding Stem Cells?, stem cell research continues to face a great deal of controversy and hesitancy from portions of the general public. Some researchers, recognizing that this controversy may deter widespread acceptance of therapeutic treatments based on stem tell technology, argue that this 'fear of the unknown' is one of the biggest obstacles impeding the practical application of stem cell research (Zakrzewski et al., 2019). To this end, these researchers argue that before stem cell technology can be widely integrated into human medical practice, more research needs to be done to fully understand the mechanisms underlying stem cell function in animal models (Ikehara, 2013). This step is viewed as unavoidable; until researchers can fully understand the nuances of how stem cells function in animals, the use of stem cell technology in humans will surely remain limited and continue to be viewed with trepidation (Ikehara, 2013).

Through animal testing, these researchers hope to validate the safety and efficacy of stem cell therapies before recreating them in human patients (Harding, Roberts & Mirochnitchenko, 2013). To this end, mice models have traditionally been used to study stem cell function in animals. However there is some debate as to whether stem cell function in mice is adequately representative of how stem cells would be expected to behave in humans (Harding, Roberts & Mirochnitchenko, 2013). It is for this reason that a growing number of studies have begun studying stem cell function in larger animals, such as pigs and dogs, which are thought to be more representative of human physiology (Harding, Roberts & Mirochnitchenko, 2013). For instance, some studies have pointed to the validity of using miniature pigs to assess the efficacy of stem cell therapies designed to repair cartilage damage (Chu, Szczodry & Bruno, 2010). This is because the structure, thickness, and weight distribution of pig cartilages systems are fairly proportional to human systems (Chu, Szczodry & Bruno, 2010).. That being said, because no single animal model can perfectly replicate all the nuances of stem cell function in humans, future stem cell research will likely aim to collect relevant information from as many different large animal models as possible (Cibelli et al., 2013). Collectively, these large animal models may be used to fill existing knowledge gaps and resolve safety concerns around stem cell research.

It should also be noted that, in a somewhat ironic twist, the use of animal models to validate the efficacy of stem cells may also predicate a reduction in the amount of animal testing needed for biomedical research (Kim, Che & Yun, 2019). Animal experimentation, while controversial, has traditionally been viewed as the more ethical alternative to testing potentially harmful drugs on human participants, making it a 'necessary evil' in the history of biomedical research (Kim, Che & Yun, 2019). While animal experimentation has contributed to increased life expectancies and decreased incidences of illness, it is not a perfect solution. For starters, there are significant differences between the physiologies of human and animal subjects, limiting the transferability of animal experimentation findings to human contexts (Kim, Che & Yun, 2019). In recent years, animal testing has also raised a number of ethical concerns regarding the unnecessary pain, distress, and harm inflicted on animal subjects (Sántha, 2020).

In light of both of these points of consideration, the European Commission has laid out a 3R strategy to 'reduce, refine, and replace' the number of animals currently involved in scientific research (Sántha, 2020). In more recent years, stem cell cultures have been put forward as a potential alternative to animal test subjects (Sántha, 2020). Due to their ability to differentiate into specific cell types, stem cells are a promising candidate for experimental drug screenings, as they may provide insight into how a particular human tissue would be affected when targeted by a drug or treatment (Tandon & Jyoti, 2012). That being said, in their current form, stem cell cultures have their fair share of limitations, including a lack of vasculature or immune system, rendering them an imperfect replacement for animal experimentation (Kim, Che & Yun, 2019).

SUBSTRATE MODIFICATION

Another challenge faced by stem cell researchers is the isolation and differentiation of patient stem cells for practical use. This is most prominently seen in the research on pluripotent stem cells which, despite having great potential in the fields of regenerative medicine, are limited by the difficulty of producing pluripotent stem cells in significant quantities (Chan, Rizwan & Yim, 2020). While a deeper discussion on pluripotent stem cells can be found in 'Chapter 6: What are prominent stem cell therapies and treatments used today?', it should be noted that conventional methods for producing pluripotent stem cells are constrained by their risk of contamination, cost, inefficiency, or the amount of time and manpower needed (Chan, Rizwan & Yim, 2020). These same constraints also limit large scale production of the desired stem cells (Chan, Rizwan & Yim, 2020). However, recent innovations have led to the emergence of new technologies which may offer promising alternatives.

Some of these emerging technologies are the result of manipulating the *substrate*: the surface used to grow new colonies of stem cells (Chan, Rizwan & Yim, 2020). One way of manipulating the substrate is by modifying its *topography*: the types of physical features present on the surface of a substrate (Reimer et al., 2016). For instance, one study was able to promote the growth of stem cells, while preserving their ability to differentiate, by using a focused beam of electrons to etch microscopic grooves and pillars onto the surface of their substrate (Ko et al., 2018). Substrates can also be manipulated through the modification of their *stiffness*: a measure of a substrate's resistance to change (Chan, Rizwan & Yim, 2020). To this end, one study was also able to promote the growth of stem cells by increasing the stiffness of their substrate, which they manipulated by adding small meshes of microscopic fibers (Maldonado et al., 2015). This same study also found that increasing the stiffness of their substrate decreased the amount of stem cells undergoing unwanted spontaneous differentiation, allowing them to better manage their batch of stem cells from differentiating into undesired cell types (Maldonado et al., 2015).

Other emerging technologies have strayed away from using these sorts of substrates, which are largely limited to a two-dimensional plane, and have instead sought to grow stem cells within a controlled three-dimensional environment (Schmidt, Lilienkampf & Bradley, 2018). While these 3D substrates are typically more complex and resource-intensive than their 2D counterparts, they also offer some key advantages. For starters, these 3D substrates are generally able to better mimic the actual microenvironment of the human body, thereby replicating the setting where stem cells would typically grow under natural conditions (Schmidt, Lilienkampf & Bradley, 2018). This is perhaps best illustrated by the study conducted by Lei & Schaffer (2013). This study sought to compare the rate of growth of stem cells in a 3D substrate relative to a 2D substrate over the course of a 15 day period (Lei & Schaffer, 2013). To this end, Lei & Schaffer found that the initial stem cell colony grew about 80-fold in the 3D substrate, while the stem cell colony in the 2D substrate only grew about 9-fold (2013).

Future studies are expected to build upon the innovations laid out above, refining the ability of substrates to support stem cell growth while maintaining their unique properties (Schmidt, Lilienkampf & Bradley, 2018). These innovations, while definitely impressive in themselves, have the potential to support major leaps in the field of stem cell research. In the long term, improved substrates may even predicate the development of large-scale stem cell production viable on a commercial scale, which would be integral to the hypothetical use of stem cell treatments by the general public (Chan, Rizwan & Yim, 2020).

IMMUNE REJECTION

Stem cell researchers have already begun to make strides towards overcoming the limitations surrounding one particularly promising therapeutic application of stem cell technology: stem cell transplants (Brunt, Weisel & Li, 2012). It should be noted that stem cell transplants are not at all a new phenomenon, and have been used for several years, especially when treating those with cancers of the blood and bone marrow (Léger & Nevill, 2004). In such cases, the transplants are needed to replace a patient's hematopoietic stem cells, which are responsible for producing new red blood cells, in the event that they become unhealthy and malignant (Léger & Nevill, 2004). To this day, stem cell transplants are considered to be a dangerous procedure which leaves the recipient vulnerable to potentially-deadly complications (Léger & Nevill, 2004). As such, they are typically reserved for patients with otherwise life-threatening diseases. This danger is especially present for the recipients of allogeneic stem cell transplants, in which stem cells are provided by a genetically-unrelated donor (Taylor, Bolton & Bradley, 2011). Complications occur when the recipient's immune system recognizes these transplanted stem cells as a foreign entity and attempts to destroy them through an immunological response (Taylor, Bolton & Bradley, 2011).

If this immunological barrier can be resolved, then allogeneic stem cell transplants may become a viable therapeutic treatment for more patients, not just those in life-threatening situations (Brunt, Weisel & Li, 2012). Allogeneic sources of stem cells may be especially attractive, if not outright necessary, for patients who are unable to receive transplants from a non-allogeneic source due to factors such as old age, stem cell degradation, disease limitations, or a lack of compatible relatives (Brunt, Weisel & Li, 2012). Furthermore, under the right conditions, stem cells could hypothetically be preemptively collected and stored for later use, allowing for their distribution to and availability in widespread communities in a manner analogous to blood banks (Brunt, Weisel & Li, 2012). While these proposals are indeed attractive, they do skirt around the underlying issue of the immunological incompatibility between stem cell transplants and their recipients.

Fortunately, while still in its infancy, some research has begun to address potential resolutions to this obstacle. It should first be acknowledged that prior to differentiating into a particular cell type, stem cells have a degree of immune privilege, meaning that they are unlikely to trigger an immune response when transplanted from a donor to a recipient (Huang et al., 2010). It is only after the stem cell differentiates, taking on a definitive form, that it loses its immune privilege and becomes susceptible to triggering an immune response from the recipient (Huang et al., 2010). A growing body of research has emerged in response to this discovery, seeking to explore whether stem cell transplants can provide any significant therapeutic benefit before they can differentiate and trigger this immune response (Brunt, Weisel & Li, 2012). Conversely, some researchers have sought to instead look into prolonging this period of immune privilege by either extending the period in which transplanted stem cells remain undifferentiated, or by extending their immune privilege to the differentiated form of the stem cell (Brunt, Weisel & Li, 2012). To this end, some researchers have proposed the use of novel gene editing technologies to 'cloak' differentiated stem cells in a layer of immune-restraining molecules, which would in theory allow them to avoid detection by the recipient's immune system (Trounson, Boyd & Boyd, 2019). There is great promise to this line of research. As future studies continue to make strides against the challenges posed by immune rejection, the once-distant prospect of extending the scope of stem cell transplants to other diseases becomes more attainable.

CELL-FREE THERAPY

On a first glance, the application of stem cells in 'cell-free therapy' may seem paradoxical. As a relatively recent advance in the field of stem cell research, cell-free therapy offers a new insight into the potential therapeutic applications of stem cells (Evangelista, Soares & Villarreal, 2019). Rather than viewing the stem cells themselves as the therapeutic agent, this approach simply views them as a potential source of therapeutic agents (Evangelista, Soares & Villarreal, 2019).

For context, it should be explained that over the past ten years, several studies have observed that mesenchymal stem cells, the type found in adult tissues, generate small secretions known as 'extracellular vesicles' (Riazifar et al., 2017). It is believed that these extracellular vesicles play an important role in mediating cell-to-cell communication (Bruno et al., 2019). For instance, after their release, extracellular vesicles have been observed transferring genetic information from their cell of origin to other stem cells they encounter while in circulation (Bruno et al., 2019). In some cases, the transmission of genetic information may cause the recipient stem cell to undergo a change in phenotype and function, effectively triggering its differentiation into a specific cell type (Bruno et al., 2019). In other cases, extracellular vesicles have been known to induce other changes in cell behaviour such as: the promotion of cell regeneration, the activation of dormant cells, and the mediation of inflammatory responses (Bruno et al., 2019). Through these actions, extracellular vesicles are able to mimic, if not outright replicate, many of the beneficial effects associated with the therapeutic applications of mesenchymal stem cells (Rani et al., 2015).

Relative to the transplantation of whole stem cells, the procedure which had been discussed in the previous section, the administration of extracellular vesicles may offer notable advantages (Rani et al., 2015). For instance, transplanted extracellular vesicles are less likely to be detected by the recipient's immune system, decreasing the risk of a potentially-fatal innate or adaptive immune system response (Rani et al., 2015). Furthermore, extracellular vesicles are also less likely to trigger the formation of tumours, a known issue associated with the transplantation of embryonic stem cells (Rani et al., 2015). There is also evidence to suggest that extracellular vesicles can be isolated and stored for extended periods of time without compromising their function, highlighting their potential future application as a therapeutic agent (Riazifar et al., 2017). As extracellular vesicles continue to attract attention from the stem cell research community, there is no telling what innovative breakthroughs may come next.

CONCLUSION

Despite its relative novelty, stem cell research has rallied a great amount of intrigue from both academic circles and the general public. As stem cell research continues to surge forward, the promise that its therapeutic applications will revolutionize medicine seems all the more tangible.

Regardless, there are still many obstacles which must be overcome before stem cell therapy can be presented as a viable option for many. Some of these obstacles have been summarized in the preceding section, alongside estimations of how they may guide the future of stem cell research. That being said, each significant breakthrough seems to raise just as many questions as its answers, leading researchers down the path of increasingly innovative advances. The future of stem cell research is bright.

References

CHAPTER 1

Bogner, A., Jouneau, P., Thollet, G., Basset, D., & Gauthier, C. (2006, July 17). *A history of scanning electron microscopy developments: Towards "wet-STEM" imaging.* Micron. https://www.sciencedirect.com/science/article/pii/S0968432806001016?via%3Dihub.

Bongso, A., & Richards, M. (2004, October 31). *History and perspective of stem cell research.* Best Practice & Research Clinical Obstetrics & Gynaecology. https://www.sciencedirect.com/science/article/pii/S1521693404001415?casa_token=Hvm9WPJPM2gAAAAA%3A7-G7Y-jcwbCdfFnq8QjCOzr7_JgQzhxVi3baOklDe0-V34Hxv2zaSv2PdpeMwm-5PLzvnVoA7kS4A#!

Ex Vivo & In Vivo Gene Therapy Techniques. Gene Therapy. (2020). https://www.thegenehome.com/how-does-gene-therapy-work/techniques.

Jacków, J., Guo, Z., Hansen, C., Abaci, H. E., Doucet, Y. S., Shin, J. U., ... Christiano, A. M. (2019, December 26). *CRISPR/Cas9-based targeted genome editing for correction of recessive DYSTROPHIC epidermolysis BULLOSA using iPS cells.* https://www.pnas.org/content/116/52/26846.

Ji, Y., Ma, Y., Chen, X., Ji, X., Gao, J., Zhang, L., ... Hu, J. (2017, August 1). *Microvesicles released from human embryonic stem cell derived-mesenchymal stem cells inhibit proliferation of leukemia cells.* Oncology Reports. https://www.spandidos-publications.com/10.3892/or.2017.5729.

Joung, J. K., & Sander, J. D. (2013). TALENs: a widely applicable technology for targeted genome editing. *Nature reviews. Molecular cell biology,* *14*(1), 49–55. https://doi.org/10.1038/nrm3486

Lee, J., Bayarsaikhan, D., Bayarsaikhan, G., Kim, J.-S., Schwarzbach, E., & Lee, B. (2020, February 12). *Recent advances in genome editing of stem cells for drug discovery and therapeutic application.* Pharmacology & Therapeutics. https://www.sciencedirect.com/science/article/pii/S0163725820300292.

Liu, D., Zheng, W., Pan, S., & Liu, Z. (2020, November 21). *Concise review: current trends on applications of stem cells in diabetic nephropathy.* Nature News. https://www.nature.com/articles/s41419-020-03206-1.

Masters, B. R. (2009, March 15). *History of the Electron Microscope in Cell Biology.* Massachusetts Institute of Technology. http://fen.bilkent.edu.tr/~physics/news/masters/ELS_HistoryEM.pdf.

Miko, M., Danišovič, L., Majidi, A., & Varga, I. (2015). Ultrastructural analysis of different human mesenchymal stem cells after in vitro expansion: a technical review. *European journal of histochemistry: EJH,* *59*(4), 2528. https://doi.org/10.4081/ejh.2015.2528

Pai-Dhungat, J. (2020, August). Invention of electron microscope. Retrieved May 17, 2021, from https://www.japi.org/x2646464/invention-of-electron-microscope

Patmanathan, S. N., Gnanasegaran, N., Lim, M. N., Husaini, R., Fakiruddin, K. S., & Zakaria, Z. (2018). CRISPR/Cas9 in Stem Cell Research: Current Application and Future Perspective. *Current stem cell research & therapy,* *13*(8), 632–644. https://doi.org/10.2174/157488X13666180613081443

Ramalingam, S., London, V., Kandavelou, K., Cebotaru, L., Guggino, W., Civin, C., & Chandrasegaran, S. (2013). Generation and genetic engineering of human induced pluripotent stem cells using designed zinc finger nucleases. *Stem cells and development, 22*(4), 595–610. https://doi.org/10.1089/scd.2012.0245

Robinson, A. L. (1986). Electron microscope inventors share Nobel physics prize. *Science, 234*, 821+. https://link.gale.com/apps/doc/A4589151/AONE?u=ocul_mcmaster&sid=AONE&xid=c8894a51

UMASS. (2018, June 11). *What is Electron Microscopy?*. University of Massachusetts Medical School. https://www.umassmed.edu/cemf/whatisem/#:~:text=Because%20of%20its%20great%20depth,determination%2C%20and%20for%20process%20control.

Vemuri, M. C. (2007). Digital Imaging of Stem Cells by Electron Microscopy. In *Stem cell assays* (pp. 22-41). Totowa, N.J.: Humana Press.

CHAPTER 2

Appelbaum, F. R. (2007). Hematopoietic-Cell Transplantation at 50. *New England Journal of Medicine, 377*(15), 4. https://redbook.streamliners.co.nz/SCT%20Hematopoietic-Cell%20Transplantation%20at%2050.pdf

CDC. (2020, August 3). *What is Epigenetics?* | CDC. Centers for Disease Control and Prevention. https://www.cdc.gov/genomics/disease/epigenetics.htm

Evans, M. (2011). Discovering pluripotency: 30 years of mouse embryonic stem cells. *Nature Reviews Molecular Cell Biology, 12*(10), 680–686. https://doi.org/10.1038/nrm3190

Jordan, C. T. (2007). The leukemic stem cell. *Best Practice & Research Clinical Haematology, 20*(1), 13–18. https://doi.org/10.1016/j.beha.2006.10.005

Li, Y., Shen, Z., Shelat, H., & Geng, Y.-J. (2013). Reprogramming somatic cells to pluripotency: A Fresh Look at Yamanaka's Model. Cell Cycle, 12(23), 3594–3598. https://doi.org/10.4161/cc.26952

Smith, A. (2010). Pluripotent stem cells: Private obsession and public expectation. *EMBO Molecular Medicine, 2*(4), 113–116. https://doi.org/10.1002/emmm.201000065

Solter, D. (2006). From teratocarcinomas to embryonic stem cells and beyond: A history of embryonic stem cell research. *Nature Reviews Genetics, 7*(4), 319–327. https://doi.org/10.1038/nrg1827

Steesman, D. P., & Kyle, R. A. (2021). James Till and Ernest McCulloch: Hematopoietic Stem Cell Discoverers. *Mayo Clinic Proceedings, 96*(3), 1. https://doi.org/10.1014/j.mayocp.2021.01.016

CHAPTER 3

Amzat, J., & Razum, O. (2014). Sociology and Health. *Medical Sociology in Africa*, 1–19. https://doi.org/10.1007/978-3-319-03986-2_1

Bhagavan, N. V., & Ha, C.-E. (2015). Chapter 22—DNA Replication, Repair, and Mutagenesis. In N. V. Bhagavan & C.-E. Ha (Eds.), *Essentials of Medical Biochemistry (Second Edition)* (pp. 401–417). Academic Press. https://doi.org/10.1016/B978-0-12-416687-5.00022-1

Biehl, J. K., & Russell, B. (2009). Introduction to Stem Cell Therapy. *The Journal of Cardiovascular Nursing, 24*(2), 98–105. https://doi.org/10.1097/JCN.0b013e318197a6a5

Bone Marrow Transplantation. (n.d.). Retrieved May 14, 2021, from https://www.hopkinsmedicine.org/health/treatment-tests-and-therapies/bone-marrow-transplantation

Definition of hematopoietic stem cell—NCI Dictionary of Cancer Terms—National Cancer Institute (nciglobal,ncienterprise). (2011, February 2). [NciAppModulePage]. https://www.cancer.gov/publications/dictionaries/cancer-terms/def/hematopoietic-stem-cell

Gratwohl, A., Pasquini, M. C., Aljurf, M., Atsuta, Y., Baldomero, H., Foeken, L., Gratwohl, M., Bouzas, L. F., Confer, D., Frauendorfer, K., Gluckman, E., Greinix, H., Horowitz, M., Iida, M., Lipton, J., Madrigal, A., Mohty, M., Noel, L., Novitzky, N., … Niederwieser, D. (2015). One million haemopoietic stem-cell transplants: A retrospective observational study. *The Lancet Haematology, 2*(3), e91–e100. https://doi.org/10.1016/S2352-3026(15)00028-9

Hägi, T. T., Laugisch, O., Ivanovic, A., & Sculean, A. (2014). Regenerative periodontal therapy. *Quintessence International (Berlin, Germany: 1985), 45*(3), 185–192. https://doi.org/10.3290/j.qi.a31203

Jang, S. E., Qiu, L., Chan, L. L., Tan, E.-K., & Zeng, L. (2020). Current Status of Stem Cell-Derived Therapies for Parkinson's Disease: From Cell Assessment and Imaging Modalities to Clinical Trials. *Frontiers in Neuroscience, 14.* https://doi.org/10.3389/fnins.2020.558532

Khaddour, K., Hana, C. K., & Mewawalla, P. (2021). Hematopoietic Stem Cell Transplantation. In *StatPearls.* StatPearls Publishing. http://www.ncbi.nlm.nih.gov/books/NBK536951/

Kolios, G., & Moodley, Y. (2013). Introduction to Stem Cells and Regenerative Medicine. *Respiration, 85*(1), 3–10. https://doi.org/10.1159/000345615

Kumar, A., Pareek, V., Faiq, M. A., Ghosh, S. K., & Kumari, C. (2019). ADULT NEUROGENESIS IN HUMANS: A Review of Basic Concepts, History, Current Research, and Clinical Implications. *Innovations in Clinical Neuroscience, 16*(5–6), 30–37.

Lin, Y., & Yelick, P. C. (2008). 75—Dental Tissue Engineering. In A. Atala, R. Lanza, J. A. Thomson, & R. M. Nerem (Eds.), *Principles of Regenerative Medicine* (pp. 1286–1297). Academic Press. https://doi.org/10.1016/B978-012369410-2.50077-2

Little, W. (2016). Chapter 19. The Sociology of the Body: Health and Medicine. *In Introduction to Sociology—2nd Canadian Edition.* BCcampus. https://opentextbc.ca/introductiontosociology2ndedition/chapter/chapter-19-the-sociology-of-the-body-health-and-medicine/

Master, Z., McLeod, M., & Mendez, I. (2007). Benefits, risks and ethical considerations in translation of stem cell research to clinical applications in Parkinson's disease. *Journal of Medical Ethics, 33*(3), 169–173. https://doi.org/10.1136/jme.2005.013169

Pack, A. R. (1998). Dental services and needs in developing countries. *International Dental Journal, 48*(3 Suppl 1), 239–247. https://doi.org/10.1111/j.1875-595x.1998.tb00712.x

Racial Minorities Face a Dearth of Stem Cell Donors. (2019, November 22). *Science in the News.* https://sitn.hms.harvard.edu/flash/2019/racial-minorities-trouble-with-stem-cell-transplants-a-dearth-of-donors/

Surgical Mission Trips | ReSurge International. (n.d.). Retrieved May 17, 2021, from https://resurge.org/surgical-mission-trips/

Zakrzewski, W., Dobrzyński, M., Szymonowicz, M., & Rybak, Z. (2019). Stem cells: Past, present, and future. *Stem Cell Research & Therapy, 10.* https://doi.org/10.1186/s13287-019-1165-5

CHAPTER 4

Ahmad, Z., Wardale, J., Brooks, R., Henson, F., Noorani, A., & Rushton, N. (2012). Exploring the application of stem cells in tendon repair and regeneration. Arthroscopy: *The Journal of Arthroscopic & Related Surgery, 28*(7), 1018-1029.

Alison, M. R., Poulsom, R., Forbes, S., & Wright, N. A. (2002). An introduction to stem cells. The *Journal of Pathology: A Journal of the Pathological Society of Great Britain and Ireland, 197*(4), 419-423.

Ebert, A. D., & Svendsen, C. N. (2010). Human stem cells and drug screening: opportunities and challenges. *Nature reviews Drug discovery, 9*(5), 367-372.

Han, Y., Li, X., Zhang, Y., Han, Y., Chang, F., & Ding, J. (2019). Mesenchymal stem cells for regenerative medicine. *Cells, 8*(8), 886.

Hoogduijn, M. J., Popp, F. C., Grohnert, A., Crop, M. J., Van Rhijn, M., Rowshani, A. T., Eggenhofer, E., Renner, P., Reinders, M.E., Rabelink, T.J. and Van Der Laan, L.J. & MISOT Study Group. (2010). Advancement of mesenchymal stem cell therapy in solid organ transplantation (MISOT). *Transplantation, 90*(2), 124-126.

Importance of Stem Cells: Stem Cells: University of Nebraska Medical Center. (n.d.). Retrieved from https://www.unmc.edu/stemcells/educational-resources/importance.html

Matsa, E., Burridge, P. W., & Wu, J. C. (2014). Human stem cells for modeling heart disease and for drug discovery. *Science translational medicine, 6*(239), 239ps6-239ps6.

Patel, D. M., Shah, J., & Srivastava, A. S. (2013). Therapeutic potential of mesenchymal stem cells in regenerative medicine. *Stem cells international, 2013*.

Rubin, L. L. (2008). Stem cells and drug discovery: the beginning of a new era?. *Cell, 132*(4), 549-552.

Xu, X. L., Yi, F., Pan, H. Z., Duan, S. L., Ding, Z. C., Yuan, G. H., Qu, J., Zhang, H.C. & Liu, G. H. (2013). Progress and prospects in stem cell therapy. *Acta Pharmacologica Sinica, 34*(6), 741-746.

Zakrzewski, W., Dobrzyński, M., Szymonowicz, M., & Rybak, Z. (2019). Stem cells: past, present, and future. *Stem cell research & therapy, 10*(1), 1-22.

CHAPTER 5

4. *The adult stem cell | stemcells. Nih. Gov.* (n.d.). Retrieved May 16, 2021, from https://stemcells.nih.gov/info/2001report/chapter4.htm#:~:-text=Adult%20stem%20cells%2C%20like%20all,long%2Dterm%20self%2Drenewal.

5. *Hematopoietic Stem Cells | stemcells.nih.gov.* (n.d.). Retrieved May 16, 2021, from https://stemcells.nih.gov/info/2001report/chapter5.htm

Boveri, T. (1892). *Sitzungsberichte Gesellschaft für Morphologie und Physiologie.* 7, 117.

Chatterjee, T., Sarkar, R., Dhot, P., Kumar, S., & Kumar, V. (2010). Adult stem cell plasticity: Dream or reality? *Medical Journal, Armed Forces India, 66*(1), 56–60. https://doi.org/10.1016/S0377-1237(10)80095-4

Clayton, E., Doupé, D. P., Klein, A. M., Winton, D. J., Simons, B. D., & Jones, P. H. (2007). A single type of progenitor cell maintains normal epidermis. *Nature, 446*(7132), 185–189. https://doi.org/10.1038/nature05574

Forgacs, G., & Newman, S. A. (2005). *Biological physics of the developing embryo*. Cambridge University Press.

Gest, H. (2004). The discovery of microorganisms by robert hooke and antoni van leeuwenhoek, fellows of the royal society. *Notes and Records of the Royal Society of London, 58*(2), 187–201. https://doi.org/10.1098/rsnr.2004.0055

Häcker, V. (1892). Die kerntheilungsvorgänge bei der mesodermund entodermbildung von cyclops. *Archiv für mikroskopische Anatomie, 39*(1), 556–581. https://doi.org/10.1007/BF02961538

Haeckel, E. (1868). *Natürliche Schöpfungsgeschichte: Gemeinverständliche wissenschaftliche Vorträge über die Entwickelungslehre im Allgemeinen und diejenige von Darwin, Göthe und Lamarck im Besonderen, über die Anwendung derselben auf den Ursprung des Menschen und andern damit zusammenhängende Gründfragen der Natur-Wissenschaft. Mit Tafeln, Holzschnitten, systematischen und genealogischen Tabellen.* Reimer.

Haeckel, E. H. P. A. (1877). *Anthropogenie* (3rd ed.). W. Engelmann.

Hooke, R. (1665). *Micrographia* . https://www.bl.uk/collection-items/micrographia-by-robert-hooke-1665
Mummery, C., van de Stolpe, A., Roelen, B. A. J., & Clevers, H. (2014). Adult stem cells: Generation of self-organizing mini-organs in a dish. *Stem Cells*, 279–290. https://doi.org/10.1016/B978-0-12-411551-4.00010-6

O'Connor, C., & Adams, J. U. (2010). 5.5 *Normal Controls on Cell Division are Lost during Cancer.* NPG Education. https://www.nature.com/scitable/ebooks/essentials-of-cell-biology-14749010/122997842/

Okita, K., & Yamanaka, S. (2011). Induced pluripotent stem cells: Opportunities and challenges. *Philosophical Transactions of the Royal Society B: Biological Sciences, 366*(1575), 2198–2207. https://doi.org/10.1098/rstb.2011.0016

Richardson, M. K. (1998). Haeckel's embryos, continued. *Science, 281*(5381), 1285–1285. https://doi.org/10.1126/science.281.5381.1285j

Schultz, M. (2008). Rudolf virchow. *Emerging Infectious Diseases, 14*(9), 1480–1481. https://doi.org/10.3201/eid1409.086672

Schwann, T. (1839). *Mikroskopische untersuchungen über die uebereinstimmung in der struktur und dem wachsthum der thiere und pflanzen.* https://wellcomecollection.org/works/bknnmj2k

Shahriyari, L., & Komarova, N. L. (2013). Symmetric vs. Asymmetric stem cell divisions: An adaptation against cancer? *PLOS ONE, 8*(10), e76195. https://doi.org/10.1371/journal.pone.0076195

Shearer, W. T., Lubin, B. H., Cairo, M. S., & Notarangelo, L. D. (2017). Cord blood banking for potential future transplantation. *Pediatrics, 140*(5). https://doi.org/10.1542/peds.2017-2695

So, W.-K., & Cheung, T. H. (2018). Molecular regulation of cellular quiescence: A perspective from adult stem cells and its niches. *Methods in Molecular Biology (Clifton, N.J.), 1686*, 1–25. https://doi.org/10.1007/978-1-4939-7371-2_1

Soares, M. J., & Varberg, K. M. (2018). Trophoblast. In *Encyclopedia of Reproduction* (2nd ed.). https://www.sciencedirect.com/science/article/pii/B9780128012383646640

Takahashi, K., & Yamanaka, S. (2006). Induction of pluripotent stem cells from mouse embryonic and adult fibroblast cultures by defined factors. *Cell, 126*(4), 663–676. https://doi.org/10.1016/j.cell.2006.07.024

Thomson, J. A., Itskovitz-Eldor, J., Shapiro, S. S., Waknitz, M. A., Swiergiel, J. J., Marshall, V. S., & Jones, J. M. (1998). Embryonic stem cell lines derived from human blastocysts. *Science, 282*(5391), 1145–1147. https://doi.org/10.1126/science.282.5391.1145

Weiss, M. L., & Troyer, D. L. (2006). Stem cells in the umbilical cord. *Stem Cell Reviews, 2*(2), 155–162. https://doi.org/10.1007/s12015-006-0022-y

Wilson, E. B. (1900). *The cell in development and inheritance.* Macmillan.

Zakrzewski, W., Dobrzyński, M., Szymonowicz, M., & Rybak, Z. (2019). Stem cells: Past, present, and future. *Stem Cell Research & Therapy, 10*(1), 68. https://doi.org/10.1186/s13287-019-1165-5

CHAPTER 6

Biehl, J. K., & Russell, B. (2009). Introduction to stem cell therapy. *The Journal of cardiovascular nursing, 24*(2), 98–105. https://doi.org/10.1097/JCN.0b013e318197a6a5

Chen, S., Du, K., & Zou, C. (2020). Current progress in stem cell therapy for type 1 diabetes mellitus. *Stem Cell Research & Therapy, 11*(1), 1-13

Krans, B. (2018, September 29). *Bone Marrow Transplant: Types, Procedure & Risks.* Healthline. https://www.healthline.com/health/bone-marrow-transplant. .

Lalla RV, Sonis ST, Peterson DE. Management of oral mucositis in patients who have cancer. *Dent Clin North Am.* 2008;52(1):61-viii. doi:10.1016/j.cden.2007.10.002

Moradi, S., Mahdizadeh, H., Šarić, T. *et al.* Research and therapy with induced pluripotent stem cells (iPSCs): social, legal, and ethical considerations. *Stem Cell Res Ther* 10, 341 (2019). https://doi.org/10.1186/s13287-019-1455-y

Pluripotent Stem Cells 101., (2021). *Boston Children's Hospital.* Boston Childrens Hospital. http://stemcell.childrenshospital.org/about-stem-cells/pluripotent-stem-cells-101/.

Pond, G. R., Lipton, J. H., & Messner, H. A. (2006). Long-term survival after blood and marrow transplantation: comparison with an age- and gender-matched normative population. Biology of blood and marrow transplantation : journal of the American Society for Blood and Marrow Transplantation, 12(4), 422–429. https://doi.org/10.1016/j.bbmt.2005.11.518

Rosa, V., Dubey, N., Islam, I., Min, K. S., & Nör, J. E. (2016). Pluripotency of stem cells from human exfoliated deciduous teeth for tissue engineering. *Stem cells international, 2016.*

Wernig, M., Zhao, J. P., Pruszak, J., Hedlund, E., Fu, D., Soldner, F., ... & Jaenisch, R. (2008). Neurons derived from reprogrammed fibroblasts functionally integrate into the fetal brain and improve symptoms of rats with Parkinson's disease. *Proceedings of the National Academy of Sciences, 105*(15), 5856-5861.

Zakrzewski, W., Dobrzyński, M., Szymonowicz, M., & Rybak, Z. (2019). Stem cells: past, present, and future. *Stem cell research & therapy, 10*(1), 1-22.

CHAPTER 7

Alison, M. R., & Islam, S. (2009). Attributes of adult stem cells. The Journal of Pathology: *A Journal of the Pathological Society of Great Britain and Ireland, 217*(2), 144-160.

Alison, M. R., Poulsom, R., Forbes, S., & Wright, N. A. (2002). An introduction to stem cells. *The Journal of Pathology: A Journal of the Pathological Society of Great Britain and Ireland, 197*(4), 419-423.

Arora, M. (2013). Cell culture media: a review. *Mater methods, 3*(175), 24.

Chen, S. S., Fitzgerald, W., Zimmerberg, J., Kleinman, H. K., & Margolis, L. (2007). Cell—cell and cell—extracellular matrix interactions regulate embryonic stem cell differentiation. *Stem cells, 25*(3), 553-561.

DeBry, R. W., & Seldin, M. F. (1996). Human/mouse homology relationships. *Genomics, 33*(3).

Ding, W., Nowakowski, G. S., Knox, T. R., Boysen, J. C., Maas, M. L., Schwager, S. M., Wu, W., Wellik, L. E., Dietz, A. B., Ghosh, A. K., Secreto, C. R., Medina, K. L., Shanafelt, T. D., Zent, C. S., Call, T. G., & Kay, N. E. (2009). Bi—directional activation between mesenchymal stem cells and CLL B—cells: implication for CLL disease progression. *British journal of haematology, 147*(4), 471-483.

Huggett, J., Dheda, K., Bustin, S., & Zumla, A. (2005). Real-time RT-PCR normalisation; strategies and considerations. *Genes & Immunity, 6*(4), 279-284.

Idelson, M., Alper, R., Obolensky, A., Ben-Shushan, E., Hemo, I., Yachi-movich-Cohen, N., Khaner, H., Smith, Y., Wiser, O., Gropp, M., Cohen, M. A., Even-Ram, S., Berman-Zaken, Y., Matzrafi, L., Rechavi, G., Banin, E., & Reubinoff, B. (2009). Directed differentiation of human embryonic stem cells into functional retinal pigment epithelium cells. *Cell stem cell, 5*(4), 396-408.

Jahan-Tigh, R. R., Ryan, C., Obermoser, G., & Schwarzenberger, K. (2012). Flow cytometry. *The Journal of investigative dermatology, 132*(10), e1.

Ramirez, J. M., Gerbal—Chaloin, S., Milhavet, O., Qiang, B., Becker, F., Assou, S., Lematire, J. M., Hamamah, S., & De Vos, J. (2011). Brief report: benchmarking human pluripotent stem cell markers during differentiation into the three germ layers unveils a striking heteroge-neity: all markers are not equal. *Stem Cells, 29*(9), 1469-1474.

Sterneckert, J. L., Reinhardt, P., & Schöler, H. R. (2014). Investigating human disease using stem cell models. *Nature Reviews Genetics, 15*(9), 625-639.

Takahashi, K., & Yamanaka, S. (2006). Induction of pluripotent stem cells from mouse embryonic and adult fibroblast cultures by defined factors. *cell, 126*(4), 663-676.

Tanabe, K., Takahashi, K., & Yamanaka, S. (2014). Induction of plurip-otency by defined factors. *Proceedings of the Japan Academy, Series B, 90*(3), 83-96.

Oh, Y., & Jang, J. (2019). Directed differentiation of pluripotent stem cells by transcription factors. *Molecules and cells, 42*(3), 200.

Rhee, W. J., & Bao, G. (2009). Simultaneous detection of mRNA and protein stem cell markers in live cells. *BMC biotechnology, 9*(1), 1-10.

Zakrzewski, W., Dobrzyński, M., Szymonowicz, M., & Rybak, Z. (2019). Stem cells: past, present, and future. *Stem cell research & therapy, 10*(1), 1-22.

Zhao, W., Ji, X., Zhang, F., Li, L., & Ma, L. (2012). Embryonic stem cell markers. *Molecules, 17*(6), 6196-6236.

CHAPTER 8

4. *The Adult Stem Cell | stemcells.nih.gov.* (n.d.). Retrieved May 15, 2021, from https://stemcells.nih.gov/info/2001report/chapter4.htm

Briggs, R., & King, T. J. (1952). Transplantation of Living Nuclei From Blastula Cells into Enucleated Frogs' Eggs. *Proceedings of the National Academy of Sciences of the United States of America, 38*(5), 455–463.

Brock, D. W. (2006). Is a consensus possible on stem cell research? Moral and political obstacles. *Journal of Medical Ethics, 32*(1), 36–42. https://doi.org/10.1136/jme.2005.013581

Charitos, I. A., Ballini, A., Cantore, S., Boccellino, M., Di Domenico, M., Borsani, E., Nocini, R., Di Cosola, M., Santacroce, L., & Bottalico, L. (2021). Stem Cells: A Historical Review about Biological, Religious, and Ethical Issues. *Stem Cells International, 2021*, e9978837. https://doi.org/10.1155/2021/9978837

Farley, M. A. (2009). Chapter 67 - Stem Cell Research: Religious Considerations. In R. Lanza, J. Gearhart, B. Hogan, D. Melton, R. Pedersen, E. D. Thomas, J. Thomson, & I. Wilmut (Eds.), *Essentials of Stem Cell Biology (Second Edition)* (pp. 609–617). Academic Press. https://doi.org/10.1016/B978-0-12-374729-7.00067-6

Findlay, J. K., Gear, M. L., Illingworth, P. J., Junk, S. M., Kay, G., Mackerras, A. H., Pope, A., Rothenfluh, H. S., & Wilton, L. (2007). *Human embryo: A biological definition. Human Reproduction, 22*(4), 905–911. https://doi.org/10.1093/humrep/del467

Gouveia, C., Huyser, C., Egli, D., & Pepper, M. S. (2020). Lessons Learned from Somatic Cell Nuclear Transfer. *International Journal of Molecular Sciences, 21*(7). https://doi.org/10.3390/ijms21072314

Häyry, M. (2018). Ethics and cloning. *British Medical Bulletin, 128*(1), 15–21. https://doi.org/10.1093/bmb/ldy031

Holm, S. (2002). Going to the Roots of the Stem Cell Controversy. *Bioethics, 16*(6), 493–507. https://doi.org/10.1111/1467-8519.00307

Jaenisch, R. (2004). Human Cloning—The Science and Ethics of Nuclear Transplantation. *New England Journal of Medicine, 351*(27), 2787–2791. https://doi.org/10.1056/NEJMp048304

Murugan, V. (2009). Embryonic Stem Cell Research: A Decade of Debate from Bush to Obama. *The Yale Journal of Biology and Medicine, 82*(3), 101–103.

Nisbet, M. C., Brossard, D., & Kroepsch, A. (2003). Framing Science: The Stem Cell Controversy in an Age of Press/Politics. *Harvard International Journal of Press/Politics, 8*(2), 36–70. https://doi.org/10.1177/1081180X02251047

Sciences (US), N. A. of, Engineering (US), N. A. of, Science, I. of M. (US) and N. R. C. (US) C. on, Engineering, & Policy, and P. (2002). Cloning: Definitions And Applications. In *Scientific and Medical Aspects of Human Reproductive Cloning*. National Academies Press (US). https://www.ncbi.nlm.nih.gov/books/NBK223960/

Tan, W., Proudfoot, C., Lillico, S. G., & Whitelaw, C. B. A. (2016). Gene targeting, genome editing: From Dolly to editors. *Transgenic Research, 25*(3), 273–287. https://doi.org/10.1007/s11248-016-9932-x

Weiss, M. L., & Troyer, D. L. (2006). Stem cells in the umbilical cord. *Stem Cell Reviews, 2*(2), 155–162. https://doi.org/10.1007/s12015-006-0022-y

CHAPTER 9

Caulfield, T., & Murdoch, B. (2019). Regulatory and policy tools to address unproven stem cell interventions in Canada: The need for action. *BMC Medical Ethics, 20*(1), 51. https://doi.org/10.1186/s12910-019-0388-4

Dresser, R. (2010). Stem Cell Research as Innovation: Expanding the Ethical and Policy Conversation. *Journal of Law, Medicine & Ethics, 38*(2), 332–341. https://doi.org/10.1111/j.1748-720X.2010.00492.x

FDA, O. of the C. (2020). FDA Warns About Stem Cell Therapies. *FDA.* https://www.fda.gov/consumers/consumer-updates/fda-warns-about-stem-cell-therapies

Holm, S. (2002). Going to the roots of the stem cell controversy. *Bioethics, 16*(6), 493–507. https://doi.org/10.1111/1467-8519.00307

Hynes, R. O., Coller, B. S., & Porteus, M. (2017). Toward Responsible Human Genome Editing. *JAMA, 317*(18), 1829–1830. https://doi.org/10.1001/jama.2017.4548

Kamenova, K. (2017). Media portrayal of stem cell research: Towards a normative model for science communication. *Asian Bioethics Review, 9*(3), 199–209. https://doi.org/10.1007/s41649-017-0026-8

Kamenova, K., & Caulfield, T. (2015). Stem cell hype: Media portrayal of therapy translation. *Science Translational Medicine, 7*(278), 278ps4-278ps4. https://doi.org/10.1126/scitranslmed.3010496

Marcon, A. R., Murdoch, B., & Caulfield, T. (2017). Fake news portrayals of stem cells and stem cell research. *Regenerative Medicine, 12*(7), 765–775. https://doi.org/10.2217/rme-2017-0060

Murdoch, B., Zarzeczny, A., & Caulfield, T. (2018). Exploiting science? A systematic analysis of complementary and alternative medicine clinic websites' marketing of stem cell therapies. *BMJ Open, 8*(2), e019414. https://doi.org/10.1136/bmjopen-2017-019414

Nisbet, M. C., Brossard, D., & Kroepsch, A. (2003). Framing Science: The Stem Cell Controversy in an Age of Press/Politics. *Harvard International Journal of Press/Politics, 8*(2), 36–70. https://doi.org/10.1177/1081180X02251047

Verlinsky, Y. (2005). Designing babies: What the future holds. *Reproductive BioMedicine Online, 10,* 24–26. https://doi.org/10.1016/S1472-6483(10)62200-6

Zarzeczny, A., & Caulfield, T. (2009). Emerging Ethical, Legal and Social Issues Associated with Stem Cell Research & and the Current Role of the Moral Status of the Embryo. *Stem Cell Reviews and Reports, 5*(2), 96–101. https://doi.org/10.1007/s12015-009-9062-4

CHAPTER 10

Bongso, A., & Richards, M. (2004). History and perspective of stem cell research. *Best practice & research Clinical obstetrics & gynaecology, 18*(6), 827-842.

Bruno, S., Chiabotto, G., Favaro, E., Deregibus, M. C., & Camussi, G. (2019). Role of extracellular vesicles in stem cell biology. *American Journal of Physiology-Cell Physiology, 317*(2), C303-C313.

Brunt, K. R., Weisel, R. D., & Li, R. K. (2012). Stem cells and regenerative medicine—future perspectives. *Canadian journal of physiology and pharmacology, 90*(3), 327-335.

Chan, S. W., Rizwan, M., & Yim, E. K. (2020). Emerging Methods for Enhancing Pluripotent Stem Cell Expansion. *Frontiers in cell and developmental biology, 8,* 70.

Chu, C. R., Szczodry, M., & Bruno, S. (2010). Animal models for cartilage regeneration and repair. *Tissue Engineering Part B: Reviews, 16*(1), 105-115.

Cibelli, J., Emborg, M. E., Prockop, D. J., Roberts, M., Schatten, G., Rao, M., ... & Mirochnitchenko, O. (2013). Strategies for improving animal models for regenerative medicine. *Cell stem cell, 12*(3), 271-274.

Evangelista, A. F., Soares, M. B. P., & Villarreal, C. F. (2019). Cell-free therapy: a neuroregenerative approach to sensory neuropathy?. *Neural regeneration research, 14*(8), 1383.

Harding, J., Roberts, R. M., & Mirochnitchenko, O. (2013). Large animal models for stem cell therapy. *Stem cell research & therapy, 4*(2), 1-9.

Huang, X. P., Sun, Z., Miyagi, Y., McDonald Kinkaid, H., Zhang, L., Weisel, R. D., & Li, R. K. (2010). Differentiation of allogeneic mesenchymal stem cells induces immunogenicity and limits their long-term benefits for myocardial repair. *Circulation, 122*(23), 2419-2429.

Ikehara, S. (2013). Grand challenges in stem cell treatments. *Frontiers in cell and developmental biology, 1,* 2.

Kim, T. W., Che, J. H., & Yun, J. W. (2019). Use of stem cells as alternative methods to animal experimentation in predictive toxicology. *Regulatory Toxicology and Pharmacology, 105,* 15-29.

Ko, J. Y., Oh, H. J., Lee, J., & Im, G. I. (2018). Nanotopographic influence on the in vitro behavior of induced pluripotent stem cells. *Tissue Engineering Part A, 24*(7-8), 595-606.

Lei, Y., & Schaffer, D. V. (2013). A fully defined and scalable 3D culture system for human pluripotent stem cell expansion and differentiation. *Proceedings of the National Academy of Sciences, 110*(52), E5039-E5048.

Léger, C. S., & Nevill, T. J. (2004). Hematopoietic stem cell transplantation: a primer for the primary care physician. *Cmaj, 170*(10), 1569-1577.

Maldonado, M., Wong, L. Y., Echeverria, C., Ico, G., Low, K., Fujimoto, T., ... & Nam, J. (2015). The effects of electrospun substrate-mediated cell colony morphology on the self-renewal of human induced pluripotent stem cells. *Biomaterials, 50*, 10-19.

Rani, S., Ryan, A. E., Griffin, M. D., & Ritter, T. (2015). Mesenchymal stem cell-derived extracellular vesicles: toward cell-free therapeutic applications. *Molecular Therapy, 23*(5), 812-823.

Reimer, A., Vasilevich, A., Hulshof, F., Viswanathan, P., Van Blitterswijk, C. A., De Boer, J., & Watt, F. M. (2016). Scalable topographies to support proliferation and Oct4 expression by human induced pluripotent stem cells. *Scientific reports, 6*(1), 1-8.

Riazifar, M., Pone, E. J., Lötvall, J., & Zhao, W. (2017). Stem cell extracellular vesicles: extended messages of regeneration. *Annual review of pharmacology and toxicology, 57,* 125-154.

Sántha, M. (2020). Biologia futura: animal testing in drug development—the past, the present and the future. *Biologia Futura,* 1-10.

Schmidt, S., Lilienkampf, A., & Bradley, M. (2018). New substrates for stem cell control. *Philosophical Transactions of the Royal Society B: Biological Sciences, 373*(1750), 20170223.

Solter, D. (2006). From teratocarcinomas to embryonic stem cells and beyond: a history of embryonic stem cell research. *Nature Reviews Genetics, 7*(4), 319-327.

Tandon, S., & Jyoti, S. (2012). Embryonic stem cells: An alternative approach to developmental toxicity testing. *Journal of pharmacy & bioallied sciences, 4*(2), 96.

Taylor, C. J., Bolton, E. M., & Bradley, J. A. (2011). Immunological considerations for embryonic and induced pluripotent stem cell banking. *Philosophical Transactions of the Royal Society B: Biological Sciences, 366*(1575), 2312-2322.

Trounson, A., Boyd, N. R., & Boyd, R. L. (2019). Toward a universal solution: editing compatibility into pluripotent stem cells. *Cell stem cell, 24*(4), 508-510.

Zakrzewski, W., Dobrzyński, M., Szymonowicz, M., & Rybak, Z. (2019). Stem cells: past, present, and future. *Stem cell research & therapy, 10*(1), 1-22.

www.ingramcontent.com/pod-product-compliance
Lightning Source LLC
Chambersburg PA
CBHW070355200326
41518CB00012B/2243